Proton Pump Inhibitors: A Balanced View

Frontiers of Gastrointestinal Research

Vol. 32

Series Editor

Choitsu Sakamoto Tokyo

Proton Pump Inhibitors: A Balanced View

Volume Editors

Tsutomu Chiba Kyoto
Peter Malfertheiner Magdeburg
Hiroshi Satoh Kyoto

28 figures, 2 in color, and 17 tables, 2013

KARGER

Basel · Freiburg · Paris · London · New York · New Delhi · Bangkok · Beijing · Tokyo · Kuala Lumpur · Singapore · Sydney

Frontiers of Gastrointestinal Research

Tsutomu Chiba
Department of Gastroenterology and Hepatology
Graduate School of Medicine
Kyoto University
Kyoto (Japan)

Peter Malfertheiner
Department of Gastroenterology, Hepatology and Infectious Diseases
Otto-von-Guericke University
Magdeburg (Germany)

Hiroshi Satoh
Department of Pharmacology and Experimental Therapeutics
Division of Pathological Sciences
Kyoto Pharmaceutical University
Kyoto (Japan)

Library of Congress Cataloging-in-Publication Data

Proton pump inhibitors (2013)
Proton pump inhibitors : a balanced view / volume editors, Tsutomu Chiba, Peter Malfertheiner, Hiroshi Satoh.
pages cm -- (Frontiers of gastrointestinal research, ISSN 0302-0665 ; vol. 32)
Includes bibliographical references and indexes.
ISBN 978-3-318-02415-9 (hard cover : alk. paper) -- ISBN 978-3-318-02416-6 (eISBN)
I. Chiba, Tsutomu, editor of compilation. II. Malfertheiner, P. (Peter), 1950- editor of compilation. III. Satoh, Hiroshi, editor of compilation. IV. Title. V. Series: Frontiers of gastrointestinal research ; v. 32. 0302-0665
[DNLM: 1. Peptic Ulcer--drug therapy. 2. Gastroesophageal Reflux--drug therapy. 3. Proton Pump Inhibitors--pharmacology. 4. Proton Pump Inhibitors--therapeutic use. 5. Risk Assessment. W1 FR946E v.32 2013 / WI 350]
RC827
616.3'32--dc23
2013020683

Bibliographic Indices. This publication is listed in bibliographic services, including Current Contents® and Index Medicus.

www.karger.com
Printed in Germany on acid-free and non-aging paper (ISO 9706) by Kraft Druck GmbH, Ettlingen
ISSN 0302–0665
e-ISSN 1622–3754
ISBN 978–3–318–02415–9
e-ISBN 978–3–318–02416–6

Contents

Preface

The discovery of proton pump inhibitors (PPIs) and their development in a variety of clinical scenarios has dramatically changed the management of acid-related diseases. The therapeutic domain of PPIs ranges from relief of symptoms to cure of mucosal lesions in the upper gastrointestinal tract. PPIs are the 'mainstay' therapy in acid-related symptoms, gastroesophageal reflux disease (GERD), peptic ulceration (PU), peptic ulcer, and bleeding. PPIs have a predominant role in prophylaxis of gastroduodenal damage from NSAIDs and aspirin, and they are used in a series of other distinct disease entities.

We had the privilege through clinical trials with PPIs to gain deep insights into the basic mechanisms of upper digestive pathologies such as GERD and PU. Those who prophesized the end of potent acid suppression in peptic ulcer disease with the discovery of *Helicobacter pylori* had to learn that PPIs are an essential ingredient for all treatment regimens aimed at the eradication of *H. pylori*.

After nearly 25 years of therapeutic experience with PPIs and hundreds of millions of patients that have benefited from PPI treatment, we fully appreciate the enormous value of these drugs, which have truly become 'stellar' in gastroenterology.

When PPIs were first introduced, their use was restricted to the Zollinger-Ellison syndrome because of fear that they might be too potent in their acid suppression. This was promptly overcome when the benefits of PPIs became apparent in all acid-related diseases. PPIs are now available as over-the-counter drugs, which is certainly the strongest indication of their safety and legitimation for wide use.

The risk of side effects with PPIs is recognized to be remarkably low. Nevertheless, some adverse effects with PPIs in selected patients demand that we remain alert and continue to learn about the consequences of long-term acid suppression.

We are grateful to all the authors that contributed their immense experience to this update on PPIs.

Tsutomu Chiba, Kyoto
Peter Malfertheiner, Magdeburg
Hiroshi Satoh, Kyoto

Chiba T, Malfertheiner P, Satoh H (eds): Proton Pump Inhibitors: A Balanced View.
Front Gastrointest Res. Basel, Karger, 2013, vol 32, pp 1–17 (DOI: 10.1159/000350624)

Discovery and Development of Proton Pump Inhibitors

Hiroshi Satoh

Department of Pharmacology and Experimental Therapeutics, Division of Pathological Sciences, Kyoto Pharmaceutical University, Kyoto, Japan

Abstract

Omeprazole (OPZ), developed by AB Hässle/Astra (Sweden), and lansoprazole (LPZ), developed by Takeda Chemical Industries Ltd. (Japan), were the world's first and second proton pump inhibitors (PPIs), respectively, approved for clinical use in humans. These PPIs have been widely used for more than 20 years in the treatment of acid-related diseases such as gastroduodenal ulcers and reflux esophagitis. In the process of discovery of the PPIs, both companies independently identified the same compound (2-[(2-pyridylmethyl)sulfinyl]-1*H*-benzimidazole, timoprazole), which had strong antisecretory and antiulcer activities. LPZ and OPZ are derivatives of timoprazole; the LPZ molecule has a trifluoroethoxy group that seems to confer strong antiulcer properties in addition to the compound's antisecretory action. For example, the antisecretory effect of LPZ in rats was approximately twice as strong as that of OPZ, but the antiulcer effects were more than 10 times stronger than those of OPZ in rat models of reflux esophagitis, gastric antral ulcers, and duodenal ulcers. Furthermore, LPZ, but not OPZ, also prevented the formation of small intestinal lesions induced by NSAIDs in rats. These effects can be explained, at least in part, by the finding that LPZ not only has antisecretory activity but also mucosal protective activity in which both capsaicin-sensitive sensory neurons and nitric oxide are involved. It has also been reported that LPZ has highly specific antibacterial effects on *Helicobacter pylori*. Both OPZ and LPZ are racemates, each having two optical isomers (S and R). Both the S-isomer of OPZ (esomeprazole) and the R-isomer of LPZ (dexlansoprazole) have been developed as second-generation PPIs with enhanced bioavailability and antisecretory activities in humans. This chapter reviews the history of the discovery and development of PPIs, and discusses the unique pharmacological properties of LPZ independent from the compound's antisecretory activity.

Peptic ulcers, such as gastric and duodenal ulcers, had long been a significant and intractable condition in humans until the discovery of the histamine H_2-receptor antagonist (H_2-RA) cimetidine in the late 1970s, which led to dramatic improvement in

The author worked for Takeda Chemical Industries, Ltd. from 1969 to 1997, and was deeply involved in the discovery and development of lansoprazole.

treatment. Thereafter, many H_2-RAs, such as ranitidine and famotidine, were developed and have since been used effectively for the treatment of peptic ulcers. H_2-RAs are very useful drugs, but they also have several drawbacks, i.e. they effectively inhibit basal gastric acid secretion at nighttime but are less effective for inhibition of acid secretion induced by various stimuli such as stress and meals during the day, they are not always effective for the treatment of gastroesophageal reflux disease (GERD) and Zollinger-Ellison syndrome, and their efficacy is reduced after repeated administration. Some of these problems were resolved by the discovery of the proton pump inhibitors (PPIs), such as omeprazole (OPZ; AB Hässle/Astra, Sweden) and lansoprazole (LPZ; Takeda Chemical Industries Ltd., Japan). The inhibitory effects of PPIs on gastric acid secretion are more prolonged than those of H_2-RAs, and the drugs have significantly improved the outcome of treatment for acid-related diseases. In particular, the efficacy for the treatment of reflux esophagitis and eradication of *Helicobacter pylori* with PPIs are significantly greater than with H_2-RAs. In this chapter, both a brief history of the discovery and development of the PPIs and the unique pharmacological properties of LPZ independent from its antisecretory action are reviewed.

The Discovery of Proton Pump Inhibitors

Pharmaceutical companies often synthesize new compounds through the modification of existing compounds and subsequent evaluation of their pharmacological activities. In some cases, different pharmaceutical companies working independently may coincidentally end up synthesizing the same compound as a new drug candidate. This was in fact the case with 2-pyridylthioacetamide and timoprazole, the discovery of which played a key role in the development of LPZ and OPZ.

Discovery of 2-Pyridylthioacetamide (AG-35, CMN 131)
During initial efforts to screen for antiulcer drugs with antisecretory activity, Takeda found that 2-pyridylthioacetamide (coded by Takeda as AG-35) and its analogs had strong antisecretory and antiulcer activities [1], and in 1969 the company applied for a Japanese patent for the compounds as novel antiulcer drugs (fig. 1) [2]. In 1971, the French pharmaceutical company Servier also reported strong antisecretory activities of thioamide derivatives including AG-35 (coded by Servier as CMN 131; fig. 1) [3]. However, these compounds were not developed as new drugs, mostly due to the demonstration of acute toxicity in animals. According to a review by Olbe et al. [4], the antisecretory activity of CMN 131 reported by Servier prompted AB Hässle to initiate research in an attempt to identify structural analogs of CMN 131 with strong antisecretory activity but minimal toxicity. Fig. 1 shows the milestones in the development of PPIs by Takeda and AB Hässle. Interestingly, both companies began their research on antisecretory and antiulcer drugs with the same compound (AG-35 and CMN-131).

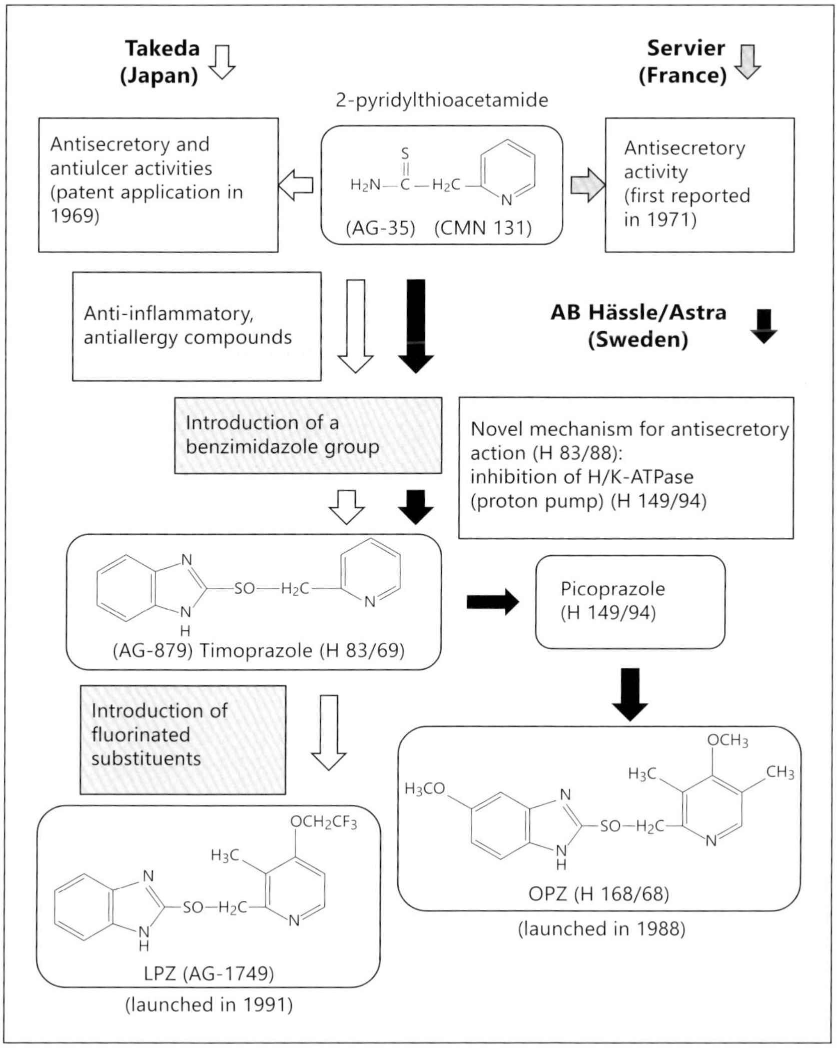

Fig. 1. Discovery and development of the PPIs LPZ and OPZ. Two compounds, 2-pyridylthioacetamide and timoprazole, played an important role in the process of discovery of LPZ (Takeda Chemical Industries Ltd.) and OPZ (AB Hässle/Astra).

Discovery of Timoprazole (AG-879, H 83/69)

The next stage in the evolution of PPIs was the identification of the compound timoprazole. AB Hässle was initially interested in CMN 131 because the drug was found to have strong antisecretory activity, although it had undesired acute toxicity. Company scientists speculated that the toxicity of CMN 131 might depend on the thioamide group in the molecule's structure. They synthesized many derivatives of CMN 131 and found that compound H 124/26, in which the thioamide group of CMN 131 was eliminated and a benzimidazole ring was incorporated, had powerful antisecretory activity but no acute toxicity. However, due to a conflicting preexisting patent, H 83/69 (a sulfoxide derivative of H 124/26, timoprazole) was selected instead for further development [4]. Subsequent long-term toxicological studies of timoprazole revealed that H 124/26 caused enlargement of the thyroid gland. Other derivatives were then screened and the compound H 149/94 (picoprazole; fig. 1) was found to have strong antisecretory activity without producing adverse effects on the thyroid/thymus. The company also reported that a derivative with a substituted benzimidazole group (coded by AB Hässle as H 83/88) inhibited acid secretion stimulated by dibutyryl-cAMP in isolated guinea pig gastric mucosa, i.e. the mode of action was quite different from that of H_2-RAs [5]. They also found that H 149/94 inhibited H^+,K^+-ATPase, which is localized in gastric parietal cells and serves as a proton pump to produce gastric acid [6]. This finding was critical for the development of the PPIs. AB Hässle eventually developed OPZ (H 168/68) as the world's first PPI (fig. 1).

From the late 1970s to early 1980s, Takeda continued to search for new antiulcer drugs with antisecretory activity. During that time ranitidine had already been approved as the second commercially available H_2-RA and, therefore, Takeda considered that additional research to develop other new H_2-RAs would not be commercially viable. However, AB Hässle's findings regarding novel mechanisms for the inhibition of acid secretion stimulated Takeda to screen for antisecretory drugs with a different mode of action from that of H_2-RAs. Through its own independent research efforts, Takeda coincidentally identified the same compound as H 83/69 – coded by Takeda as AG-879 – from a series of newly synthesized compounds that were screened for anti-inflammatory and antiallergy effects, and it was found to have strong antisecretory and antiulcer activity in rats. However, it was subsequently learned that this was the same compound as timoprazole.

Introduction of Fluorinated Substituents to Timoprazole

Takeda randomly screened more than 700 compounds in an attempt to identify antisecretory compounds with new basic chemical structures that differed from timoprazole, but Takeda eventually recognized that the basic structure of timoprazole was essential for PPI activity. Takeda investigated various modifications of timoprazole and eventually found that introduction of fluorinated substituents, such as a trifluoroethoxy group, to timoprazole (AG-879) markedly improved the antiulcer properties of the compound [7]. While human gastric ulcers are often observed in the antral area of the stomach, drug development research for this condition had been hampered by the lack of a suitable animal model for gastric antral ulcers; however, we reported that indomethacin (IND) given before or after a 1-hour feeding in rats fasted previously for 24 h caused ulcers in the antral area of the stomach [8]. The effects of fluorinated compounds were then examined in this antral ulcer model. As shown in table 1, fluorinated compounds, including LPZ, exhibited markedly stronger antiulcer activities compared with nonfluorinated reference compounds in this model [9]. The effect of LPZ was further examined in a mepirizole-induced duodenal ulcer model in

Table 1. Effect of introduction of fluorinated substituents to 2-[(2-pyridylmethyl)sulfinyl]-1*H*-benzimidazoles on IND-induced gastric antral ulcers in refed rats

Compound No.	R	ID_{50} values, mg/kg (p.o.)
(1)	CH_3	6.3
(2) LPZ (AG-1749)	CF_3	0.9
(3)	CH_2CH_3	5.1
(4)	CF_2CF_2H	<1.0
(5)	CF_2CF_3	0.8

Chemical structures of 2-[(2-pyridylmethyl)sulfinyl]-1*H*-benzimidazoles are shown in fig. 2a. Test compounds were administered orally to rats after a 24-hour fast. IND (30 mg/kg) was administered s.c. 30 min later. The animals were given chow pellets for 1 h after IND. At 5 h after the administration of IND, the animals were sacrificed and the lesion area of the gastric antrum was measured. The ID_{50} values were calculated graphically from the dose-inhibition relationships for 6–7 rats. Details are provided in [9].

rats, and it was found that the antiulcer activity of LPZ was about 10 times stronger than that of a nonfluorinated reference compound (fig. 2). These results suggest that the introduction of fluorinated substituents substantially increases the antiulcer activities of nonfluorinated 2-[(2-pyridylmethyl)sulfinyl]-1*H*-benzimidazoles. Subsequent to these findings, Takeda applied for a patent for 2-[(2-pyridylmethyl) sulfinyl]-1*H*-benzimidazoles containing fluorinated substituents as antiulcer drugs in 1984 [9], and ultimately LPZ (coded as AG-1749) was selected as a candidate antiulcer agent (fig. 1). Both LPZ and OPZ are derivatives of timoprazole, and LPZ has antisecretory properties similar to those of OPZ [10–12]; however, LPZ contains a trifluoroethoxy group which seems to provide antiulcer activity superior to that of OPZ, as well as other unique pharmacological activities, as discussed below.

Antisecretory Mechanism of Proton Pump Inhibitors

There are three types of receptors on the basolateral membrane of parietal cells: gastrin, acetylcholine, and histamine receptors. Histamine stimulates acid secretion by increasing intracellular c-AMP, and both gastrin and acetylcholine stimulate acid secretion by increasing intracellular Ca^{++}. As mentioned above, PPIs inhibit H^+,K^+-ATPase, which is localized in gastric parietal cells (tubulovesicles and intracellular canaliculi) and acts as a proton pump to produce gastric acid. Thus, PPIs inhibit all acid secretion caused by various stimuli, while H_2-RAs selectively inhibit acid secretion caused by histamine. LPZ and OPZ are taken into acid-secreting parietal cells where they covalently bind to active H^+,K^+-ATPase on intracellular canaliculi, resulting in long-lasting inhibitory effects (fig. 3). Therefore, a characteristic of these compounds is that they exert an inhibitory effect on acid secretion even when their con-

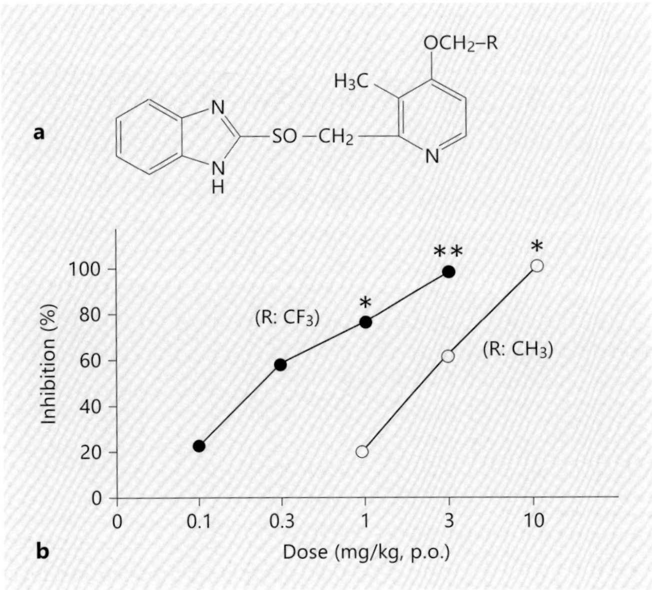

Fig. 2. Effect of introduction of a fluorinated moiety on the pharmacological activity of 2-[(2-pyridylmethyl)sulfinyl]-1*H*-benzimidazoles. **a** Chemical structure of 2-[(2-pyridylmethyl)sulfinyl]-1*H*-benzimidazoles. R: CF_3 (LPZ), CH_3 (reference compound). **b** Effect of LPZ (CF_3) and its reference compound (CH_3) on mepirizole-induced duodenal ulcers in rats. The compounds or vehicle were administered p.o. 30 min before mepirizole (200 mg/kg, s.c.). The duodenal lesions were examined 24 h later. Data show the mean inhibition values for 10–12 rats. * $p < 0.05$, ** $p < 0.01$ vs. vehicle. Details are provided in [13].

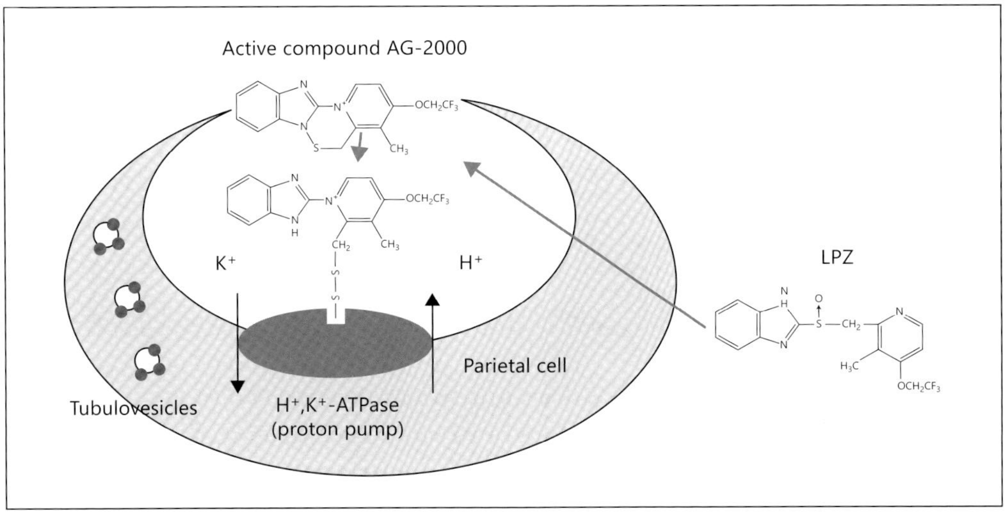

Fig. 3. Mechanism of proton pump inhibition by LPZ. Acid secretion stimuli cause tubulovesicles containing H^+,K^+-ATPase to fuse with the luminal membrane; H^+,K^+-ATPase is activated on the luminal membrane and acid is secreted. After LPZ reaches the parietal cell acidic region in the stomach, H^+ is added and converted to an activated substance that binds to H^+,K^+-ATPase through an S-S bond, thereby inhibiting acid secretion. Details are provided in [10, 12]. This figure was kindly provided by Dr. N. Inatomi, Takeda Pharmaceutical Co. Ltd.

Table 2. Effects of LPZ and OPZ on gastric acid secretion and the formation of various types of gastrointestinal lesions in rats

	LPZ ID_{50} mg/kg	OPZ ID_{50} mg/kg	Ratio of ID_{50} values (OPZ/LPZ)
Antisecretory activity			
Basal acid secretion	3.6	8.5	2.4
Histamine stimulated	1.6	3.3	2.1
Antiulcer activity			
Gastric corpus lesions			
W-I stress	2.4	7.0	2.9
IND, fasted rat	3.7	9.7	2.6
Reflux esophagitis			
Pylorus ligated	0.7	13.7	19.6
Gastric antral lesions			
IND, refed rat	0.9	21.5	23.9
Duodenal lesions			
Mepirizole	0.3	3.0	10.0

The ID_{50} values were calculated from the dose-inhibition relationships for 6–12 rats by the least squares method or graphically (gastric antral lesions). W-I stress: Water-immersion stress. Details are provided in [9, 11, 13].

centrations in blood decrease, unlike H_2-RAs. Studies on their mechanisms of action have suggested that both LPZ and OPZ do not exert a direct effect, but rather are converted to active metabolites under acidic conditions, and it is these metabolites that are responsible for inhibiting H^+,K^+-ATPase (fig. 3) [4, 10–12]. The active metabolites responsible for the effects were later identified as cyclic sulfenamides (AG-2000 for LPZ). Both LPZ and OPZ are prodrugs, and they are thought to be converted to their respective active metabolites through a unique intramolecular rearrangement reaction.

Pharmacological Properties of Lansoprazole Unrelated to Antisecretory Activity

Protective Effects of Lansoprazole on Gastrointestinal Mucosa

Both LPZ and OPZ inhibited the gastric secretion and formation of gastrointestinal lesions induced by various stimuli in rats in a dose-dependent manner (table 2). Both LPZ and OPZ inhibited basal and histamine-stimulated acid secretion. The ID_{50} values for LPZ were 3.6 mg/kg (basal) and 1.6 mg/kg (histamine-stimulated), respectively [11]. The ratios of ID_{50} values for OPZ and LPZ (OPZ/LPZ) were 2.4 and 2.1, respectively (table 2). Both LPZ and OPZ inhibited the formation of gastric corpus lesions induced by water-immersion stress or IND, and the ratios (OPZ/LPZ) of ID_{50} values were 2.9 and 2.6, respectively. In addition, the ID_{50} values of LPZ for corpus lesions (2.4 mg/kg, p.o. and 3.7 mg/kg, p.o.) are very close to those for antisecretory effects (3.6 and 1.6 mg/kg). Therefore, LPZ seems to inhibit corpus lesions mainly by inhibiting acid secretion. In contrast, the ID_{50} ratios (OPZ/LPZ) for reflux gastritis, gastric antral ulcers, and duodenal ulcers are very high, i.e. 19.6, 23.9, and 10.0, respectively (table 2). Furthermore, the ID_{50} values for LPZ in these models are 0.7, 0.9, and 0.3 mg/kg, respectively. The values were roughly one quarter to one twelfth of the ID_{50} value for inhibition of basal acid secretion (3.6 mg/kg), suggesting that LPZ has acid-independent protective effects on the gastrointestinal mucosa. Furthermore, LPZ prevented the formation of gastric lesions induced by absolute ethanol and acidified taurocholic acid [11, 14].

Collectively, these results strongly suggested that LPZ protects the gastric mucosa by a mechanism distinct from its antisecretory action. We found that the effects of LPZ on gastric lesions were markedly inhibited by pharmacological denervation of capsaicin-sensitive sensory neurons (CSSNs) or pretreatment with Nω-nitro-L-arginine methyl ester (L-NAME), a selective inhibitor of nitric oxide (NO) synthesis, suggesting that both CSSNs and NO are involved in the protective mechanism of LPZ [14]. In addition, we found that LPZ applied on the gastric mucosa using a Lucite® chamber in rats increased both NO production and blood flow in the mucosa, and that the effects were prevented by pretreatment with L-NAME [14]. We also found that LPZ increased bicarbonate secretion in the duodenum in rats, and that LPZ enhanced acid-stimulated bicarbonate secretion [15]. The response to LPZ was also abolished by prior denervation of CSSNs. The unique antiulcer action of LPZ is summarized in fig. 4.

Effects of Lansoprazole on Small Intestinal Lesions Induced by NSAIDs

Recent advances in endoscopy, such as the development of capsule endoscopy and double-balloon (push) endoscopy, have led to the finding that small intestinal lesions induced by NSAIDs, including aspirin, in humans are more common and more severe than previously thought [16, 17]. Gastric and duodenal lesions caused by NSAIDs can be treated with antisecretory agents such as PPIs or H_2-RAs [18]; however, the effects of these drugs on NSAID-induced small intestinal lesions are still not fully understood.

Effects of Single Administration

We recently examined the effects of antisecretory drugs on small intestinal lesions induced by NSAIDs in rats, and found that both H_2-RAs (cimetidine, ranitidine, and famotidine) and PPIs (OPZ and rabeprazole) at high doses exacerbated the lesions, although LPZ inhibited the formation of lesions dose dependently (fig. 5) [19]. The inhibitory effect of LPZ was observed not only in the normal rats but also in arthritic rats, and the effect was prevented by denervation of CSSNs and pretreatment with L-NAME [19], suggesting that CSSNs and NO mediate the protective effect of LPZ. Kuroda et al. [20] reported that LPZ had an inhibitory effect on IND-induced small in-

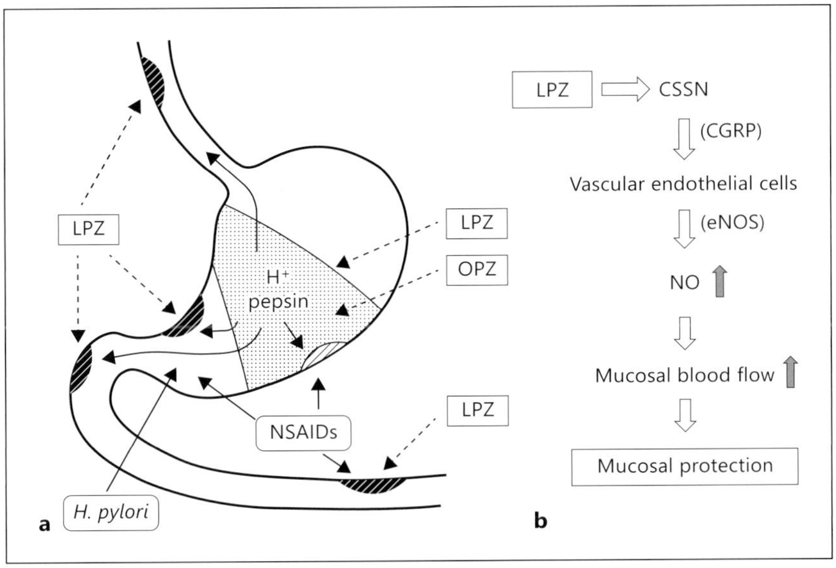

Fig. 4. Schematic diagram of the sites of mucosal protection by LPZ and OPZ, and possible mechanisms for protection of the gastrointestinal mucosa by LPZ. **a** LPZ showed stronger antiulcer activities than OPZ in the esophagus, gastric antrum, duodenum, and small intestine. **b** LPZ protects the mucosa by increasing mucosal blood flow via both CSSNs and NO. CGRP = Calcitonin gene-related peptide; eNOS = endothelial NO synthase.

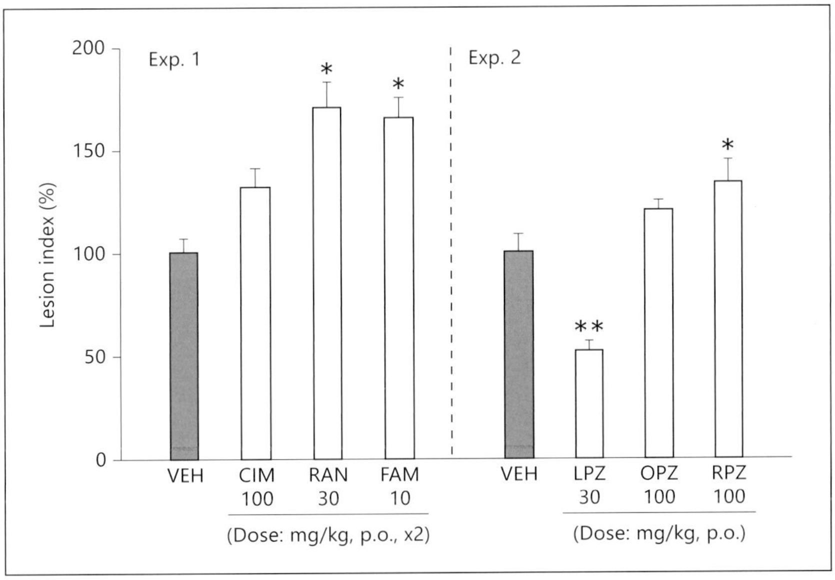

Fig. 5. Effects of H_2-RAs and PPIs on diclofenac-induced small intestinal lesions in rats. Diclofenac (10 mg/kg) was administered without fasting, and intestinal lesions were measured 24 h after diclofenac administration. H_2-RAs were administered twice, 30 min before and 6 h after diclofenac. PPIs were administered 30 min before diclofenac. VEH = Vehicle; CIM = cimetidine; RAN = ranitidine; FAM = famotidine; RPZ = rabeprazole. Data represent means ± SE for 7 rats. * $p < 0.05$, ** $p < 0.01$ vs. VEH. Details are provided in [19].

testinal lesions in rats, and suggested that LPZ prevented intestinal lesions by exerting anti-inflammatory and antioxidant effects since LPZ inhibited the increase of thiobarbituric acid-reactive substances, myeloperoxidase activity, and cytokine-induced neutrophil chemoattractant-1 induced by IND. Yoda et al. [21] reported that LPZ, but not OPZ, prevented intestinal lesions induced by IND in rats, and they hypothesized that LPZ protected the mucosa by inhibiting inducible NO synthase expression via upregulation of heme oxygenase-1/CO production in the mucosa.

Effects of Repeated Administration

Wallace et al. [22] recently reported that PPIs (OPZ and LPZ) given twice a day for 9 days in rats exacerbated small intestinal lesions induced by naproxen given twice daily over the last 4 days. They suggested that PPIs exacerbated the intestinal lesions via dysbiosis, i.e. a significant reduction of jejunal Actinobacteria and *Bifidobacteria* spp. The mechanism by which PPIs caused dysbiosis was not precisely determined in their study [22]; however, it has been reported that long-term administration of PPIs in humans increases the number of bacteria not only in the stomach, but also in the small intestine [23]. Therefore, it is possible that repeated administration of PPIs in rats causes dysbiosis in the small intestine indirectly via pH changes in the stomach. We examined the effects of repeated administration of antisecretory drugs on intestinal lesions induced by IND in rats [Satoh et al., unpubl. observation], but we could not confirm the results of Wallace et al. [22], probably due to differences in the experimental models and procedures used in the two studies. Thus, it is difficult to draw firm conclusions regarding the effects of long-term and repeated administration of PPIs on NSAID-induced intestinal lesions. Nonetheless, this is an important issue because many patients take PPIs to prevent upper gastrointestinal side effects associated with NSAID use, and the incidence of intestinal lesions in patients who take both PPIs and NSAIDs is high (50–70%) [16, 17].

Effect of Lansoprazole on Helicobacter pylori

Subsequent to the findings of Warren and Marshall [24] that *H. pylori* plays a role in the pathogenesis of active gastritis in humans, numerous studies have been carried out on the role of *H. pylori* in gastroduodenal ulcers, relapse of ulcers, and gastric cancers. Takeda's researchers demonstrated that long-term infection with *H. pylori* resulted in adenocarcinoma in the stomach in Mongolian gerbils [25]. This was the first published report of an animal model of gastric cancer induced by *H. pylori*.

Today, PPIs are commonly used to eradicate *H. pylori*. However, in the 1980s, it is likely that few researchers and clinicians expected PPIs to have antibacterial effects on *H. pylori*. In the late 1980s we examined the effects of many antiulcer drugs including LPZ on the growth of *H. pylori* in vitro, and discovered serendipitously that LPZ prevented the growth of *H. pylori* [26, 27]. As shown in table 3, LPZ inhibited the growth of *H. pylori* with an MIC_{50} of 6.25 μg/ml, and the effect was bactericidal rather than bacteriostatic [26, 28]. We then examined the effect of LPZ on the growth of other

Table 3. Effects of various antiulcer agents against *H. pylori*

Agent	MIC (range), μg/ml	MIC_{50}	MIC_{90}
Bismuth citrate	6.25–25.0	6.25	12.5
LPZ (AG-1749)	3.13–12.5	6.25	6.25
AG-2000[1]	0.78–3.13	3.13	3.13
OPZ	12.5–50.0	25.0	25.0
Cimetidine	800 to >800	>800	>800
Ranitidine	>800	>800	>800
Famotidine	>800	>800	>800

MICs were determined using 17 strains of *H. pylori* according to the agar dilution method on *Brucella* agar with a bacterial suspension of about 5×10^7 CFU/ml. Details are provided in [26]. [1] Acid converted active form of LPZ.

bacterial strains, reasoning that a nonselective antibacterial action would complicate the use of LPZ as an antiulcer drug. Surprisingly, the antibacterial effect of LPZ was highly specific for *H. pylori*; LPZ did not inhibit the growth of other bacterial strains such as *Escherichia coli, Lactobacillus acidophilus, Staphylococcus aureus,* and *Streptococcus pneumoniae,* even at concentrations as high as 100 μg/ml [26]. OPZ also had an antibacterial effect on *H. pylori,* but the activity was about one fourth that of LPZ.

As mentioned previously, LPZ inhibits H^+,K^+-ATPase by binding to the SH group of the enzyme after being transformed into its cyclic sulfenamide form (AG-2000) under acidic conditions [10]. As shown in table 3, AG-2000 had an antibacterial action against *H. pylori* and its effect was greater than that of LPZ [26]. We examined the antibacterial mechanism of LPZ, OPZ, and AG-2000, and interestingly found that both of the PPIs and AG-2000 inhibited urease activity of *H. pylori* by binding to the enzyme's SH groups [29]. The finding that PPIs inhibit urease activity by binding to SH groups, as was seen with the inhibition of H^+,K^+-ATPase, was unexpected. The specific mode of action of LPZ on urease activity may explain, at least in part, why LPZ did not show any antibacterial effects on other bacterial strains. However, we later found that inhibition of urease activity may not be related to the antibacterial action of LPZ on *H. pylori* [30]. Therefore, additional studies are needed to clarify the inhibitory mechanism of LPZ on *H. pylori.* Although the PPIs have antibacterial effects on *H. pylori* in vitro, it is unlikely that they have such an effect in patients since the PPIs are used clinically in enteric-coated forms due to the instability of the compounds in acidic environments. PPIs are expected to increase intragastric pH, which creates favorable conditions for the eradication of *H. pylori* by antibiotics. Therefore, PPIs are currently used with antibiotics and antimicrobial agents to eradicate *H. pylori.* However, repeated administration of PPIs increases the intragastric pH, which may enhance dissolution of the enteric-coating materials in the stomach. Under these conditions of elevated intragastric pH, PPIs might be effective for the eradication of *H. pylori* in the stomach.

Development of Proton Pump Inhibitors

When a novel compound with suitable pharmacological activity is identified, much additional research is required to ensure that the agent can be safely used in patients. Due to unexpected toxic effects, many drug candidates that appear to be promising in the early stages of development are ultimately abandoned. As described below, long-term toxicity studies in rats revealed undesirable toxic effects of both OPZ and LPZ.

Long-Term Toxicity Studies
Omeprazole. The development of OPZ as an antisecretory drug progressed considerably earlier than that of LPZ. Antisecretory studies on OPZ in humans were begun in 1981 and the first report of the drug's efficacy for inhibition of acid secretion was published in 1983 [4]. However, during this time span, 2-year carcinogenicity studies in rats indicated that high doses of OPZ were associated with enterochromaffin-like (ECL) cell carcinoids in the stomach, and as a result clinical studies in humans were halted in 1984. Subsequently, the pathogenesis of OPZ-induced carcinoids was intensively studied, and it was eventually shown that high-doses of OPZ produced ECL cell hyperplasia (and carcinoids) in rats by inducing hypergastrinemia. This was further supported by the finding that hyperplasia in response to high-dose OPZ did not occur in rats subjected to resection of the gastric antrum [31]. Furthermore, ECL cell carcinoids were also produced in lifelong studies of rats administered high doses of an H_2-RA (ranitidine) [32]. The results of these animal studies prompted the resumption of clinical studies on OPZ, the details of which have been reviewed by Olbe et al. [4]. After resolving the carcinoid issue, OPZ was launched in Europe in 1988 as the world's first approved PPI (fig. 1). PPIs and H_2-RAs have been used for more than 20 years in humans for the treatment of acid-related diseases, and to date, there have been no clinical reports of drug-related ECL cell carcinoids. The commercial and clinical success of PPIs would not have been possible had it not been for the efforts of various investigators to clarify the mechanism of carcinoid induction in animals.

Lansoprazole. To confirm the safety of oral administration of LPZ, approximately 80 toxicity studies, including acute, subacute, chronic, reproductive, genotoxicity, and carcinogenicity studies, were conducted [33]. During a New Drug Application (NDA) review in the USA, testicular interstitial cell tumors observed in a 2-year oral carcinogenicity study in rats became an issue. Takeda demonstrated that human interstitial cells were less sensitive to luteinizing hormone than those of rats, and that a related metabolite found in rats did not exist in humans. Based on these data, the NDA for LPZ was approved by the US Food and Drug Administration (FDA). The article [34] describing the toxicological studies received the most distinguished paper award in 1995 from the US Society of Toxicology. Following the completion of numerous basic science and clinical studies, LPZ was launched in France in 1991 as the world's second commercially available PPI (fig. 1).

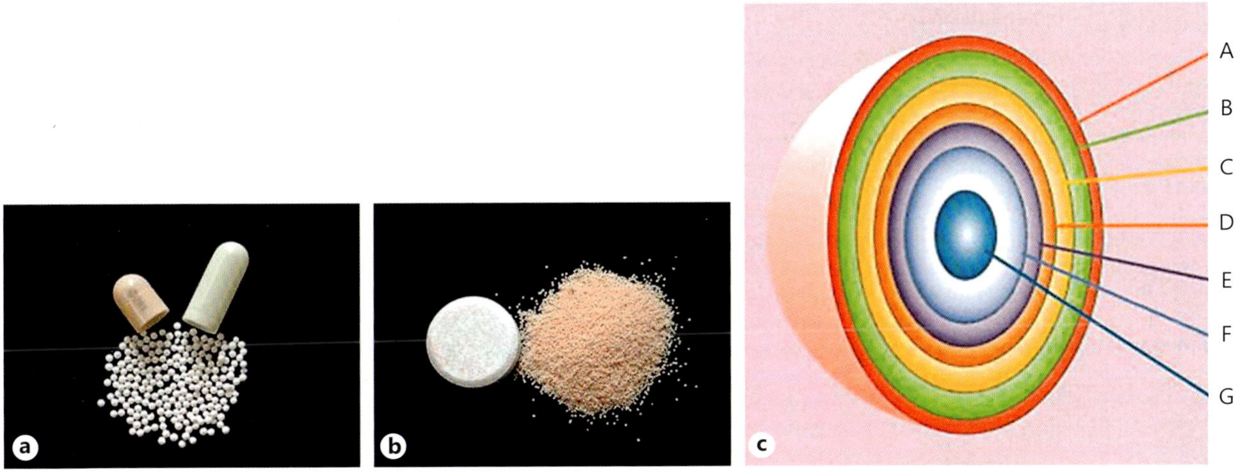

Fig. 6. Enteric-coated granules in capsule (**a**), enteric-coated microgranules in oral disintegrating tablet (**b**), and cross-section of enteric-coated microgranules (**c**). A: over-coating layer; B: third enteric coating layer containing macrogol 6,000; C: second coating layer containing triethyl citrate; D: first enteric-coating layer containing macrogol 6,000; E: intermediate layer; F: active compound layer; G: core. Details are provided in [35]. These photos were kindly provided by Dr. T. Tabata, Takeda Pharmaceutical Co. Ltd.

Development of Orally Disintegrating Lansoprazole Tablets

Since LPZ, like OPZ, is unstable under acidic conditions, an enteric-coated formulation was developed to prevent degradation of the active ingredient by gastric acid. Enteric dosage forms consist of two types of formulations: enteric-coated tablets and enteric-coated granules. It is well known that enteric-coated granules have better absorption properties than enteric-coated tablets. Phase 1 pharmacokinetic studies showed that enteric-coated LPZ tablets were absorbed more slowly in some subjects. Therefore, enteric-coated LPZ granules were further developed, and this formulation is still in use today.

To reduce the size of the tablets and make them easier for patients to ingest, Takeda developed orally disintegrating tablets made from enteric-coated granules. The tablets were designed to disintegrate within 30 s in the oral cavity, and to be easier to swallow than capsules while maintaining bioequivalence. The diameter of the enteric-coated granules (1 mm) used in the capsules was as small as possible, and the softness of the enteric-coated granules was maintained. The tablets that were ultimately developed are composed of enteric-coated microgranules and outer inactive granules so as to maintain good absorption properties, and have been shown to be bioequivalent to the capsule formulation. The enteric-coated microgranules consist of 7 layers to prevent damage to the enteric-coating layer from compression, to improve stability, and to mask the bitter taste (fig. 6) [35].

Development of an Injectable Formulation of Lansoprazole
Clinical research interest subsequently focused on the use of intravenous PPIs as a potential treatment for maintaining high gastric pH to inhibit and prevent the recurrence of upper gastrointestinal bleeding due to physical and/or psychological stress in cases of trauma or major surgery. Approximately 60 toxicity studies were conducted in animals to evaluate the safety of an injectable version of LPZ, and the formulation was ultimately approved for clinical treatment of upper gastrointestinal bleeding [33].

Development of Single Enantiomers of Omeprazole and Lansoprazole (Esomeprazole, Dexlansoprazole)
Clinical research indicated that there was large variation in the efficacy of OPZ among patients [4]. The difference in the responses was especially pronounced between slow and rapid metabolizers of OPZ. This finding prompted additional studies to elucidate the mechanism underlying the variation in response, and it was found that polymorphism of the drug-metabolizing enzyme (CYP2C19) was responsible.

It is well known that hepatic drug metabolism differs between optical isomers. Both OPZ and LPZ are racemates, each having two optical isomer forms (S and R). Optical isomers may significantly differ from each other with respect to pharmacokinetic and pharmacodynamic properties and molecular interactions. Thus, one isomer may offer significant pharmacokinetic and therapeutic advantages as compared with the other isomer or the racemic mixture. The bioavailability and antisecretory effects of racemic OPZ, S-OPZ, and R-OPZ were compared in humans, and S-OPZ (esomeprazole) was identified as the most suitable enantiomer since it exhibited superior bioavailability and oral potency in inhibiting gastric acid secretion as compared with R-OPZ or racemic OPZ in humans [36]. Conversely, Takeda selected the R-isomer of LPZ (dexlansoprazole) for a dual delayed-release formulation, based on the finding that S-LPZ is metabolized more rapidly than R-LPZ [37].

Development of a Potassium-Competitive Acid Blocker (TAK-438)
Although PPIs are the current mainstay of therapy for acid-related diseases, several areas for improvement still remain: (1) as PPIs are acid labile and are administered in acid-protective formulations, such as enteric-coated tablets, their onset of action and efficacy are greatly affected by gastric emptying; (2) their onset of action is slow, with 3–5 days of treatment needed to achieve full efficacy; (3) there is large variation in efficacy among patients because of CYP2C19 metabolism, and (4) PPIs are unable to continuously control acid secretion for a period of 24 h, even when administered twice daily. For improved intragastric pH control, several strategies for therapy other than the use of PPIs have been investigated. A new class of acid suppressants known as potassium-competitive acid blockers, or acid pump antagonists, that inhibit gastric H^+,K^+-ATPase in a K^+-competitive and reversible

manner hold promise for improving the treatment of acid-related diseases. Takeda is now developing a novel potassium-competitive acid blockers (TAK-438) as a third-generation antisecretory drug [38, 39]. TAK-438 can accumulate in gastric glands in larger quantities than LPZ, and its clearance from the gastric glands as well as dissociation from H^+,K^+-ATPase is very slow; thus, TAK-438 shows greater potency and a longer duration of antisecretory action than LPZ in animals and humans [38, 40].

Comments

Over the past two decades of our research on the pharmacological effects of LPZ, several important questions have arisen. Why does LPZ have much stronger antiulcer effects in the esophagus, gastric antrum, and duodenum than OPZ, in contrast with the effect on the gastric corpus? Why is LPZ able to prevent NSAID-induced small intestinal lesions, while OPZ tends to increase the lesions? Why does LPZ have a selective antibacterial effect on *H. pylori* and almost no effect on other strains of bacteria? Some of these questions have been at least partially elucidated, i.e. LPZ and OPZ are derivatives of timoprazole and have antisecretory properties similar to each other [10–12], but LPZ has a trifluoroethoxy group which seems to confer mucosal protective activity that is superior to or qualitatively different from OPZ. Although the antisecretory action is the most important pharmacological effect of LPZ, other pharmacological properties of LPZ might also be effective for the treatment of acid-related gastrointestinal diseases, as well as acid-independent diseases such as NSAID-induced intestinal lesions.

Takeda developed LPZ independently from AB Hässle/Astra, but during the development process, Takeda scientists learned much from the outstanding research published by AB Hässle and we respectfully acknowledge the company's valuable contributions in introducing OPZ as the world's first PPI approved for clinical use in humans.

Acknowledgements

The author is greatly indebted to Drs. A. Nohara, N. Inatomi, and T. Tabata of Takeda Pharmaceutical Co. Ltd., Osaka, Japan, and Prof. K. Takeuchi, Kyoto Pharmaceutical University, Kyoto, Japan, for valuable discussion and suggestions. The author also thanks Drs. K. Kubo, H. Nagaya, T. Iwahi and the colleagues of Takeda Pharmaceutical Co. Ltd., Osaka, Japan, for their excellent collaborative efforts in the discovery and development of LPZ.

References

1 Kanno M, Narumi S, Hirata T, Gomaibashi K, Kanai Y, Nohara A, Sanno Y: Inhibition of the gastric secretion in the rat by thioacetamide derivatives. J Takeda Res Labs 1973;32:181–187.

2 Kanai Y, Nohara A, Honda H, Kanno M, Sanno Y: inventors, Takeda Chemical Ind. Ltd., assignee. Process for producing thioamide derivatives. Japanese patent application, 1969 July (Japanese Patent Notification 1972;47–41902).

3 Malen CE, Danree BH: New thiocarboxamides derivatives with specific gastric antisecretory properties. J Med Chem 1971;14:244–246.

4 Olbe L, Carlsson E, Lindberg P: A proton-pump inhibitor expedition: the case histories of omeprazole and esomeprazole. Nat Rev Drug Discov 2003;2: 132–139.

5 Sjöstrand SE, Ryberg B, Olbe L: Stimulation and inhibition of acid secretion in the isolated guinea pig gastric mucosa. Acta Physiol Scand 1978;184(suppl): 181–185.

6 Fellenius E, Berglindh T, Sachs G, Olbe L, Elander B, Sjöstrand SE, Wallmark B: Substituted benzimidazoles inhibit gastric acid secretion by blocking (H^+ + K^+)ATPase. Nature 1981;290:159–161.

7 Kubo K, Oda K, Kaneko T, Satoh H, Nohara A: Synthesis of 2-[[(4-fluoroalkoxy-2-pyridyl) methyl] sulfinyl]-1H-benzimidazoles as antiulcer agents. Chem Pharm Bull 1990;38:2853–2858.

8 Satoh H, Inada I, Hirata T, Maki Y: Indometacin produces gastric antral ulcers in the refed rat. Gastroenterology 1981;81:719–725.

9 Nohara A, Maki Y: inventors, Takeda Chemical Ind. Ltd., assignee. 2-[2-Pyridylmethylthio-(sulfinyl)-benzimidazoles. United States patent US4628098. December 1986.

10 Nagaya H, Satoh H, Kubo K, Maki Y: Possible mechanism for the inhibition of gastric (H^+ + K^+)-adenosine triphosphatase by the proton pump inhibitor AG-1749. J Pharmacol Exp Ther 1989;248:799–805.

11 Satoh H, Inatomi N, Nagaya H, Inada I, Nohara A, Nakamura N, Maki Y: Antisecretory and antiulcer activities of a novel proton pump inhibitor AG-1749 in dogs and rats. J Pharmacol Exp Ther 1989;248: 806–815.

12 Nagaya H, Satoh H, Maki Y: Possible mechanism for the inhibition of acid formation by the proton pump inhibitor AG-1749 in isolated canine parietal cells. J Pharmacol Exp Ther 1990;252:1289–1295.

13 Inatomi N, Nagaya H, Takami K, Shino A, Satoh H: Effects of a proton pump inhibitor, AG-1749 (lansoprazole), on reflux esophagitis and experimental ulcers in rats. Jpn J Pharmacol 1991;55:437–451.

14 Murakami I, Satoh H, Asano S, Maeda R: Role of capsaicin-sensitive sensory neurons and nitric oxide in the protective effect of lansoprazole, a proton pump inhibitor, on the gastric mucosa in rats. Jpn J Pharmacol 1996;72:137–147.

15 Inada I, Satoh H: Capsaicin-sensitive sensory neurons are involved in bicarbonate secretion induced by lansoprazole, a proton pump inhibitor, in rats. Dig Dis Sci 1996;41:785–790.

16 Graham DY, Opekum AR, Willingham FF, Qureshi WA: Visible small-intestinal mucosal injury in chronic NSAID users. Clin Gastroenterol Hepatol 2005;3: 55–59.

17 Maiden L, Thjodleifsson B, Theodors A, Gonzalez J, Bjarnason I: A quantitative analysis of NSAID-induced small bowel pathology by capsule enteroscopy. Gastroenterology 2005;128:1172–1178.

18 Scarpignato C, Hunt RH: Nonsteroidal antiinflammatory drug-related injury to the gastrointestinal tract: clinical picture, pathogenesis, and prevention. Gastroenterol Clin North Am 2010;39:433–464.

19 Satoh H, Amagase K, Takeuchi K: Exacerbation of nonsteroidal anti-inflammatory drug-induced small intestinal lesions by antisecretory drugs in rats: the role of intestinal motility. J Pharmacol Exp Ther 2012;343:270–277.

20 Kuroda M, Yoshida N, Ichikawa H, Takagi T, Okuda T, Naito Y, Okanoue T, Yoshikawa T: Lansoprazole, a proton pump inhibitor, reduces the severity of indomethacin-induced rat enteritis. Int J Mol Med 2006;17:89–93.

21 Yoda Y, Amagase K, Kato S, Tokioka S, Murano M, Kakimoto K, Nishio H, Umegaki E, Takeuchi K, Higuchi K: Prevention by lansoprazole, a proton pump inhibitor, of indomethacin-induced small intestinal ulceration in rats through induction of heme oxygenase-1. J Physiol Pharmacol 2010;61:287–294.

22 Wallace JL, Syer S, Denou E, de Palma G, Vong L, McKnight W, Jury J, Bolla M, Bercik P, Collins SM, Verdu E, Ongini E: Proton pump inhibitors exacerbate NSAID-induced small intestinal injury by inducing dysbiosis. Gastroenterology 2011;141:1314–1322.

23 Lombardo L, Foti M, Ruggia O, Chiecchio A: Increased incidence of small intestinal bacterial overgrowth during proton pump inhibitor therapy. Clin Gastroenterol Hepatol 2010;8:504–508.

24 Warren JR, Marshall BJ: Unidentified curved bacilli on gastric epithelium in active chronic gastritis. Lancet 1983;i:1273–1275.

25 Watanabe T, Tada M, Nagai H, Sasaki S, Nakao M: *Helicobacter pylori* infection induced gastric cancer in Mongolian gerbils. Gastroenterology 1998;115: 642–648.

26 Iwahi T, Satoh H, Nakao M, Iwasaki T, Yamazaki T, Kubo K, Tamura T, Imada A: Lansoprazole, a novel benzimidazole proton pump inhibitor, and its related compounds have selective activity against *Helicobacter pylori*. Antimicrob Agents Chemother 1991; 35:490–496.
27 Iwahi T, Satoh H: inventors, Takeda Chemical Ind. Ltd., assignee. Selective antibacterial agent against campyrobacter. United States patent US5013743. May 1991.
28 Nakao M, Malfertheiner P: Growth inhibitory and bactericidal activities of lansoprazole compared with those of omeprazole and pantoprazole against *Helicobacter pylori*. Helicobacter 1998;3:21–27.
29 Nagata K, Satoh H, Iwahi T, Shimoyama T, Tamura T: Potent inhibitory action of the gastric proton pump inhibitor lansoprazole against urease activity of *Helicobacter pylori*: unique action selective for *H. pylori* cells. Antimicrob Agents Chemother 1993;37: 769–774.
30 Nagata K, Takagi E, Tsuda M, Nakazawa T, Satoh H, Nakao M, Okamura H, Tamura T: Inhibitory action of lansoprazole and its analogs against *Helicobacter pylori*: inhibition of growth is not related to inhibition of urease. Antimicrob Agents Chemother 1995; 39:567–570.
31 Larsson H, Carlsson E, Mattsson H, Lundell L, Sundler F, Sundell G, Wallmark B, Watanabe T, Håkanson R: Plasma gastrin and gastric enterochromaffin-like cell activation and proliferation. Studies with omeprazole and ranitidine in intact and antrectomized rats. Gastroenterology 1986;90:391–399.
32 Havu N, Mattsson H, Ekman L, Carlsson E: Enterochromaffin-like cell carcinoids in the rat gastric mucosa following long-term administration of ranitidine. Digestion 1990;45:189–195.
33 Inatomi N, Kubo K: Lansoprazole; in: History of Innovative Research and Drug Discovery at Takeda. Osaka, Takeda Pharmaceutical Co. Ltd., 2011, pp 68–75.
34 Fort FL, Miyajima H, Ando T, Suzuki T, Yamamoto M, Hamashima T, Sato S, Kitazaki T, Mahony MC, Hodgen GD: Mechanism for species-specific induction of Leydig cell tumors in rats by lansoprazole. Fundam Appl Toxicol 1995;26:191–202.
35 Tabata T: Formulation development using enteric coated granules – evolution from capsules to oral disintegrating tablets. Pharm Tech Japan 2008;24: 157–162.
36 Andersson T, Röhss K, Bredberg E, Hassan-Alin M: Pharmacokinetics and pharmacodynamics of esomeprazole, the S-isomer of omeprazole. Aliment Pharmacol Ther 2001;15:1563–1569.
37 Emerson CR, Marzella N: Dexlansoprazole: a proton pump inhibitor with a dual delayed-release system. Clin Ther 2010;32:1578–1596.
38 Hori Y, Matsukawa J, Takeuchi T, Nishida H, Kajino M, Inatomi N: A study comparing the antisecretory effect of TAK-438, a novel potassium-competitive acid blocker, with lansoprazole in animals. J Pharmacol Exp Ther 2011;337:797–804.
39 Shin JM, Inatomi N, Munson K, Strugatsky D, Tokhtaeva E, Vagin O, Sachs G: Characterization of a novel potassium-competitive acid blocker of the gastric H,K-ATPase, 1-[5-(2-fluorophenyl)-1-(pyridin-3-ylsulfonyl)-1H-pyrrol-3-yl]-N-methylmethanamine monofumarate (TAK-438). J Pharmacol Exp Ther 2011;339:412–420.
40 Jenkins R, Sakurai Y, Nishimura A, Okamoto H, Hibberd M, Jenkins H, Taubel J: Gastric acid suppression effect of TAK-438, a potassium-competitive acid blocker following ascending multiple doses in healthy subjects. Gut 2012;61:A201.

Hiroshi Satoh, DVM, PhD
Department of Pharmacology and Experimental Therapeutics
Division of Pathological Sciences, Kyoto Pharmaceutical University
Misasagi, Yamashina, Kyoto 607-8414 (Japan)
E-Mail h_satoh@mb.kyoto-phu.ac.jp

Chiba T, Malfertheiner P, Satoh H (eds): Proton Pump Inhibitors: A Balanced View.
Front Gastrointest Res. Basel, Karger, 2013, vol 32, pp 18–33 (DOI: 10.1159/000350626)

Pharmacokinetics, Pharmacodynamics, and Pharmacogenetics of Proton Pump Inhibitors

Takahisa Furuta[a] · Mitsushige Sugimoto[b] · Naohito Shirai[c]

[a]Center for Clinical Research and [b]First Department of Medicine, Hamamatsu University School of Medicine, and [c]Department of Gastroenterology, Koseiren Enshu Hospital, Hamamatsu, Japan

Abstract

Proton pump inhibitors (PPIs), such as omeprazole, lansoprazole, rabeprazole, esomeprazole, and pantoprazole, are metabolized by cytochrome P450 2C19 (CYP2C19) in the liver. There are genetic differences in the activity of this enzyme. The genotypes of CYP2C19 are classified into four groups: ultrarapid metabolizers; rapid metabolizers (RMs), which are also known as homozygous extensive metabolizers; intermediate metabolizers (IMs), which are also called heterozygous extensive metabolizer, and poor metabolizers (PMs). The pharmacokinetics and pharmacodynamics of PPIs differ among the different CYP2C19 genotype groups. Plasma PPI levels and intragastric pH during PPI treatment in the RM group are the lowest, those in the IM group come next, and those in the PM group are the highest of the three groups. These CYP2C19 genotype-dependent differences in the pharmacokinetics and pharmacodynamics of PPIs are reflective of the cure rates for gastroesophageal reflux disease and *Helicobacter pylori* infection by PPI-based therapies. The CYP2C19 genotyping test is a useful tool for the optimal and safe treatment with a PPI. PPIs sometimes interact with other drugs through CYP2C19 and its potent acid inhibition.

Proton pump inhibitors (PPIs), such as omeprazole, lansoprazole, pantoprazole, rabeprazole, and esomeprazole, are now clinically used as potent gastric acid inhibitors. Omeprazole, lansoprazole, and pantoprazole are first-generation PPIs, while rabeprazole and esomeprazole are second-generation PPIs. After absorption from the small intestine, PPIs are transferred to the gastric parietal cells via systemic circulation and bind to the proton pump (H^+/K^+-ATPase) irreversibly disturbing the function of the proton pump and thus resulting in potent acid inhibition. The major indications of PPIs are acid-related diseases, such as peptic ulcer, gastroesophageal reflux disease (GERD), and Zollinger-Ellison syndrome. PPIs are also used for the eradication of *Helicobacter pylori* with antimicrobial agents.

PPIs are substitutes for benzimidazole and undergo hepatic metabolism by the cytochrome P450 system. The metabolic pathways of PPIs are demonstrated in figure 1. The principal enzyme involved in the metabolism of PPIs is cytochrome P450 2C19 (CYP2C19). Because the metabolites of PPIs (such as 5-hydroxy-omeprazole and omeprazole sulfone in the case of omeprazole) have no acid-inhibitory effect, the therapeutic effect of PPIs depends on their plasma levels, which are determined by the activity of CYP2C19. There are genotypic differences in the activity of CYP2C19 [1]. Moreover, there is sometimes a drug-drug interaction of PPIs with other substrates of CYP2C19. Potent acid inhibition also affects the absorption of other drugs. Herein, we describe and discuss the pharmacogenomics, pharmacokinetics, and pharmacodynamics of PPIs. Drug interactions associated with PPIs are also described.

Effects of CYP2C19 Polymorphism on the Pharmacokinetics and Pharmacodynamics of Proton Pump Inhibitors

Genotype and Phenotypes of CYP2C19

Various genetic mutations involved in the CYP2C19 polymorphism have been discovered from ethnically different populations (http://www.imm.ki.se/CYPalleles/cyp2c19.htm). The polymorphism of CYP2C19 is now classified into four genotype groups: ultrarapid metabolizers (UMs); rapid metabolizers (RMs), which are also known as homozygous extensive metabolizers; intermediate metabolizers (IMs), which are also called heterozygous extensive metabolizers, and poor metabolizers (PMs) by the combination of *CYP2C19*2* of exon 5, *CYP2C19*3* of exon 4, and *CYP2C19*17* in most cases [2–6]. In RMs, neither of the two alleles have mutations and the enzyme can be generated from both of the nonmutated alleles. In IMs, the one allele has the mutation in the coding region of CYP2C19. However, the other allele has no mutation and normal enzymes can be generated from this allele. In PMs, both of the alleles have mutations in the CYP2C19 genes; therefore, normal enzymes cannot be generated from any of the two mutated alleles, resulting in the deficiency of the enzyme activity. On the other hand, UMs are known to possess the *17 allele and have marked elevated levels of enzyme activity.

There are interethnic differences in the allele frequencies [3, 7–11]. *CYP2C19*3* is rare in Caucasians. On the other hand, the incidence of *CYP2C19*17* in Asians is not so high in comparison with Caucasians (table 1).

Influences of CYP2C19 Polymorphism on the Pharmacokinetics and Pharmacodynamics of Proton Pump Inhibitors

When 20 mg of omeprazole, 30 mg of lansoprazole, or 20 mg of rabeprazole are given, plasma concentrations of the three PPIs differ among the three different *CYP2C19* genotype groups (RM, IM, and PM; fig. 2) [12]. Plasma omeprazole levels in the PM

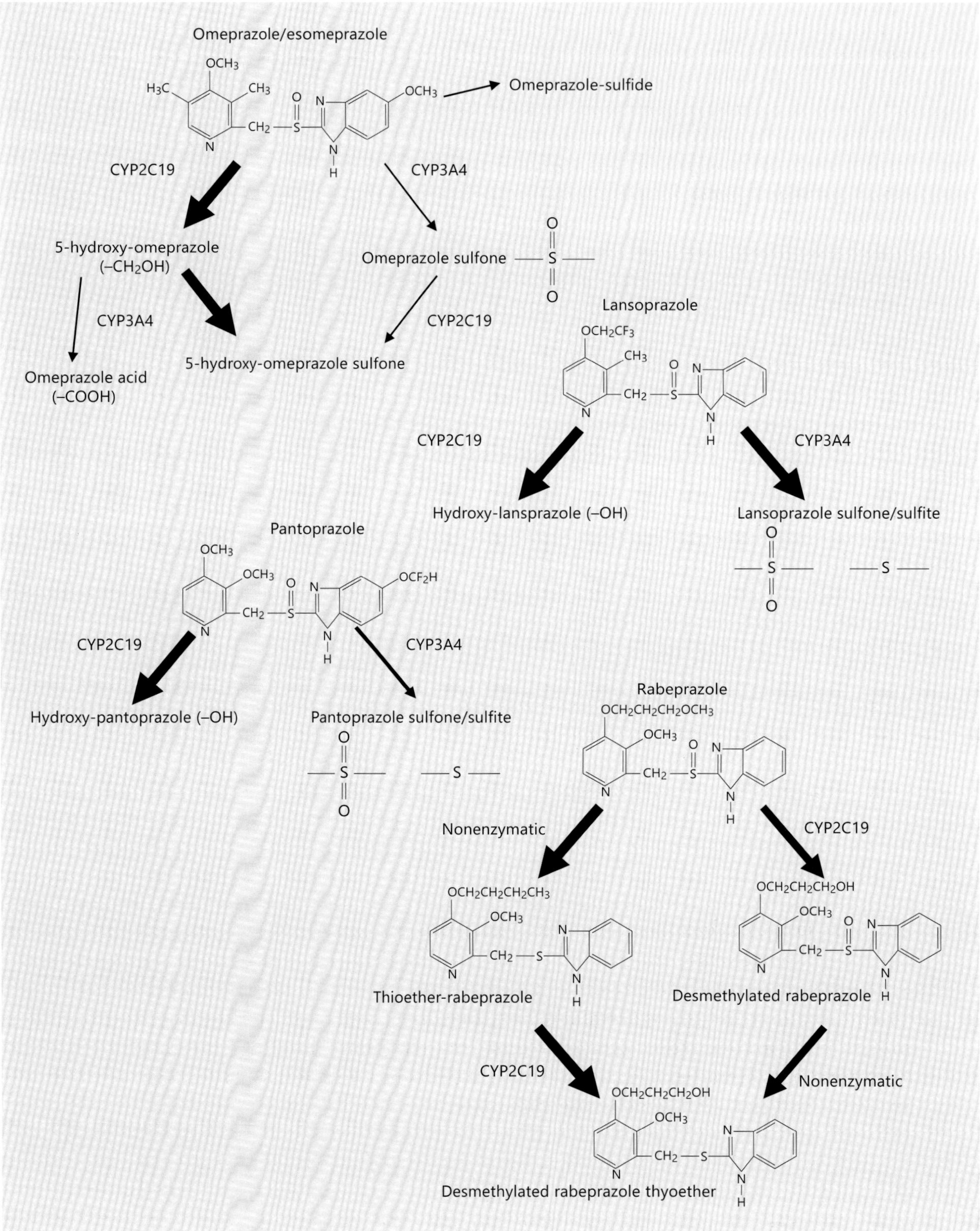
Omeprazole/esomeprazole
Omeprazole-sulfide
CYP2C19
CYP3A4
5-hydroxy-omeprazole
(–CH2OH)
Omeprazole sulfone
CYP3A4
CYP2C19
Omeprazole acid
(–COOH)
5-hydroxy-omeprazole sulfone
Lansoprazole
CYP2C19
CYP3A4
Hydroxy-lansprazole (–OH)
Lansoprazole sulfone/sulfite
Pantoprazole
CYP2C19
CYP3A4
Hydroxy-pantoprazole (–OH)
Pantoprazole sulfone/sulfite
Rabeprazole
Nonenzymatic
CYP2C19
Thioether-rabeprazole
Desmethylated rabeprazole
CYP2C19
Nonenzymatic
Desmethylated rabeprazole thyoether

Table 1. Frequencies of CYP2C19 phenotypes and alleles in different populations

Ethnicity	Phenotype frequency				Allele frequency				Ref.
	UM	RM	IM	PM	*1	*2	*3	*17	
White	0.042	0.69	0.19	0.028	0.66	0.12	NE	0.18	[7]
Caucasian	NE	0.69	0.28	0.028	0.83	0.17	0.00	NE	[6]
Caucasian	0.042	0.620	0.301	0.036	0.593	0.181	0.006	0.220	[8]
Ashkenazi Jewish	0.026	0.68	0.22	0.018	0.70	0.12	NE	0.13	[7]
Japanese	0.015	0.366	0.453	0.188	0.579	0.279	0.128	0.013	[9]
Japanese	NE	0.349	0.463	0.188	0.58	0.29	0.13	NE	[10]
Chinese	NE	0.34	0.41	0.26	0.54	0.40	0.06	NE	[3]
Africans	NE	0.68	0.281	0.037	0.823	0.173	0.004	NE	[11]
African American	0.033	0.69	0.15	0.067	0.63	0.12	NE	0.19	[7]
Hispanic	0.023	0.74	0.2	0.0087	0.75	0.10	NE	0.10	[7]

NE = Not examined.

Table 2. CYP2C19 and the means of $AUC_{0-24\,h}$ (ng·h/ml) of PPIs

Regimen	RM	IM	PM	RM:IM:PM ratio	Ref.
OPZ 20 mg day 1	421	1,403	5,109	1:3.3:12.2	[12]
OPZ 20 mg day 1	524	1,096	5,607	1:2.1:10.7	[13]
OPZ 20 mg day 8	1,057	2,418	7,153	1:2.3:6.8	[13]
RPZ 20 mg day 1	696	1,609	2,329	1:2.3:3.3	[13]
RPZ 20 mg day 8	464	1,398	2,437	1:3.0:5.3	[13]
LPZ 30 mg day 8	1,980	4,775	10,663	1:2.4:5.4	[14]
EPZ 20 mg day 5	3,280	7,280	9,190	1:2.2:2.8	[15]
EPZ 40 mg day 7	16,208	21,038	24,059	1:1.3:1.5	[16]

OPZ = Omeprazole; LPZ = lansoprazole; RPZ = rabeprazole; EPZ = esomeprazole.

group are sustained for a long time after dosing. The mean values for the area under the plasma concentration-time curves (AUC) of omeprazole in the PM group is about 13 times higher than those of the RM group (table 2). Similar *CYP2C19* genotype-dependent differences in the plasma levels of PPIs have been observed in other PPIs such as lansoprazole, rabeprazole, and esomeprazole (table 2) [12–16]. Although there are some differences in individual PPIs, the plasma levels of all PPIs are generally influenced by *CYP2C19* genotypic differences.

Fig. 1. Metabolism of omeprazole, esomeprazole, lansoprazole, pantoprazole, and rabeprazole (RPZ) in relation to cytochrome P450 (CYP) isoenzymes. Weight of arrows indicate the relative contribution of different enzyme pathways. CYP2C19 is involved in the metabolism of all PPIs.

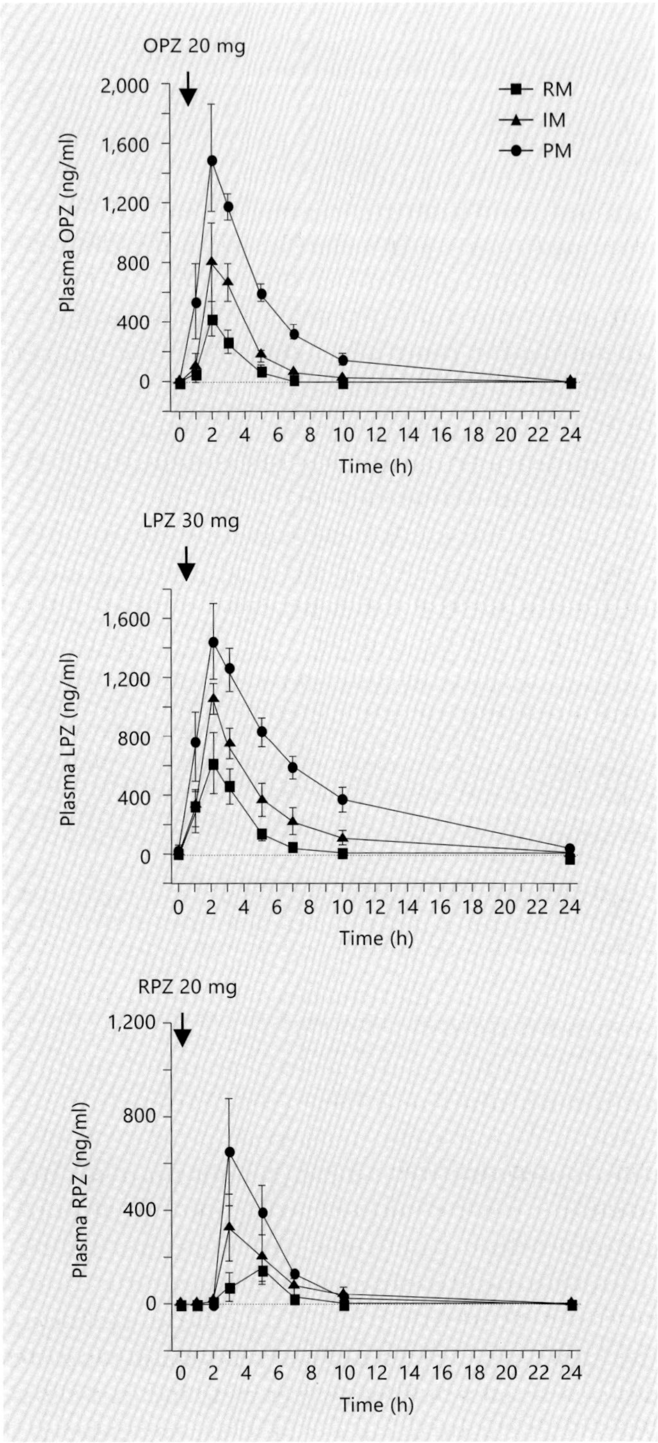

Fig. 2. Mean (±SE) plasma levels of 20 mg of omeprazole (OPZ), 30 mg of lansoprazole (LPZ), and 20 mg of rabeprazole (RPZ) as a function of CYP2C19 genotype for 8 days. Plasma concentrations of OPZ, LPZ, and RPZ are highest in the PM group, intermediate in the IM group, and lowest in the RM group.

Omeprazole, a representative PPI, is mainly metabolized by CYP2C19 to 5-hydroxy-omeprazole as noted above, and then metabolized to 5-hydroxy-omeprazole sulfone (fig. 1). Omeprazole is also in part metabolized by CYP3A4 to omeprazole sulfone, and then metabolized by CYP2C19 to 5-hydroxy-omeprazole sulfone. Plasma levels of omeprazole and their metabolites are shown in figure 3. Because hydroxylation of omeprazole by CYP2C19 cannot be performed in PMs of CYP2C19, most of omeprazole is metabolized by CYP3A4 to omeprazole sulfone. However, omeprazole sulfone cannot be metabolized by CYP2C19 to 5-hydroxy-omeprazole sulfone, and the higher plasma level of omeprazole sulfone is sustained for a long time in PMs. On the other hand, in RMs and IMs of CYP2C19, omeprazole is mainly metabolized to 5-hydroxy-omeprazole, and then metabolized to 5-hydroxy-omeprazole sulfone. In PMs, lower levels of 5-hydroxy-omeprazole are observed. Hydroxylation of omeprazole in PM is mediated by CYP3A4.

Differences in plasma levels of PPIs among different *CYP2C19* genotype groups result in the differences in intragastric pH profiles when a PPI (i.e. omeprazole 20 mg, lansoprazole 30 mg, or rabeprazole 20 mg) is administered (fig. 4). The mean 24-hour intragastric pH level is lowest in the RM group, intermediate in the IM group, and highest in the PM group [12]. The acid inhibition achieved by once daily dosing of omeprazole or lansoprazole at the usual standard doses in the RM group is likely to become insufficient under the so-called standardized dosing scheme [12, 17]. However, in cases of the second-generation PPIs, such as esomeprazole and rabeprazole, although acid inhibition appears to differ among CYP2C19 genotype groups, the relatively sufficient acid inhibition can be attained in RMs in comparison with the first-generation PPIs.

This CYP2C19 RM genotype-related insufficient acid inhibition results in the insufficient clinical outcomes of the PPI-based therapies. When omeprazole or lansoprazole at the standard doses were dosed once daily, cure rates of reflux esophagitis were reported to depend on *CYP2C19* genotypes (table 3) [18–25]. Similarly, the eradication rates of *H. pylori* infection by the PPI-based therapies, such as triple therapy with a PPI, amoxicillin, and clarithromycin, depend on CYP2C19 genotypes (table 4) [26–32]. These CYP2C19 genotype-dependent differences in the clinical outcomes seem smaller and are sometimes statistically not significant in the second-generation PPIs in comparison with the first-generation PPIs [24, 25, 33–35] (tables 3, 4).

To overcome CYP2C19 RM-related insufficient acid inhibition, several attempts have been reported. Because profound acid inhibition in PMs is considered to be derived from higher plasma PPI concentrations for a longer period, the modification of dosing schedules of PPIs is expected to induce profound acid inhibition for RMs. In our previous reports, 4 times daily dosing of lansoprazole 30 mg or rabeprazole 10 mg could attain sufficient acid inhibition [14, 36]. Frequent dosing of a PPI enables a PPI to sustain the plasma levels higher than the sufficient level for a longer time (fig. 5). Dosing of esomeprazole 10 mg 4 times daily or 20 mg twice daily was also reported to

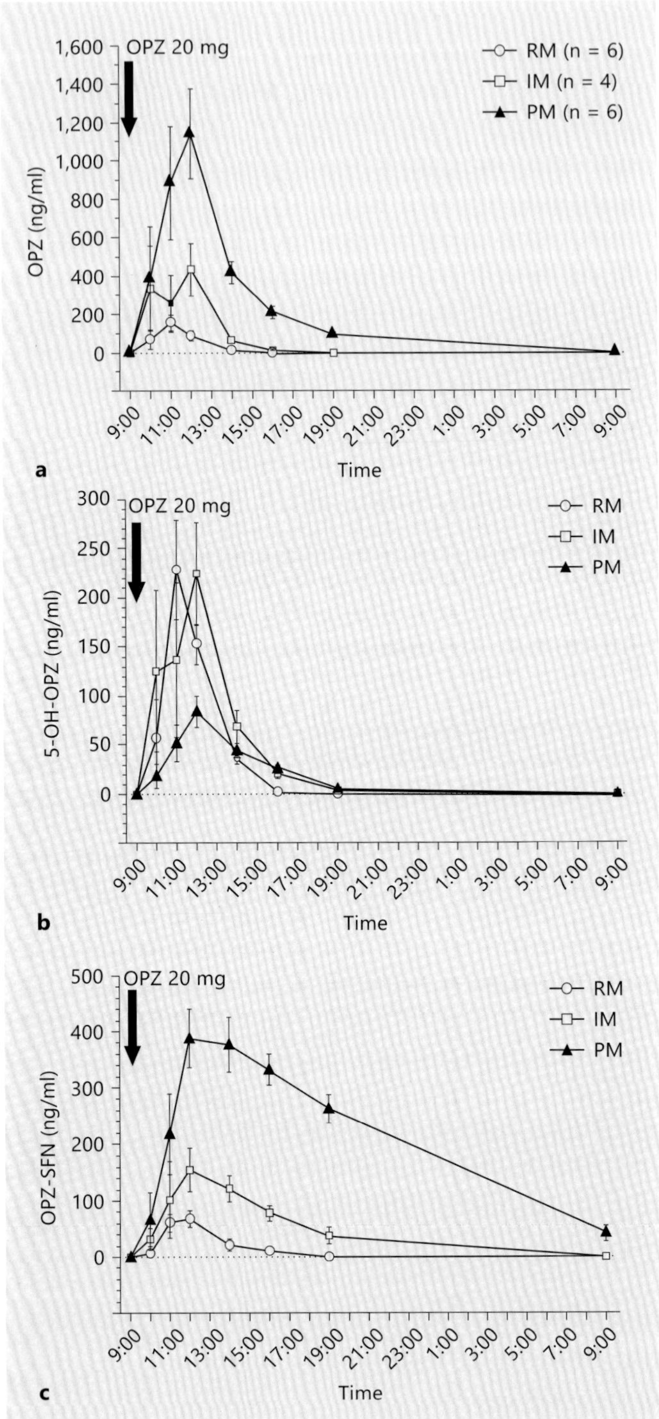

Fig. 3. Mean (±SE) plasma levels of omeprazole (OPZ) (**a**), 5-hydroxyomeprazole (5-OH-OPZ) (**b**), and omeprazole sulfone (OPZ-SFN) (**c**) as a function of the CYP2C19 genotype. **a** Plasma concentration of OPZ was highest in the PM group, intermediate in the IM group, and lowest in the RM group. **b** The peak concentration of 5-OH-OPZ was observed first in the RM group. The plasma concentration of OH-OPZ was lowest in the PM group. **c** The plasma concentration of OPZ-SFN was highest in the PM group, intermediate in the IM group, and lowest in the RM group. RM = Homozygous extensive metabolizer; IM = heterozygous extensive metabolizer; PM = poor metabolizer.

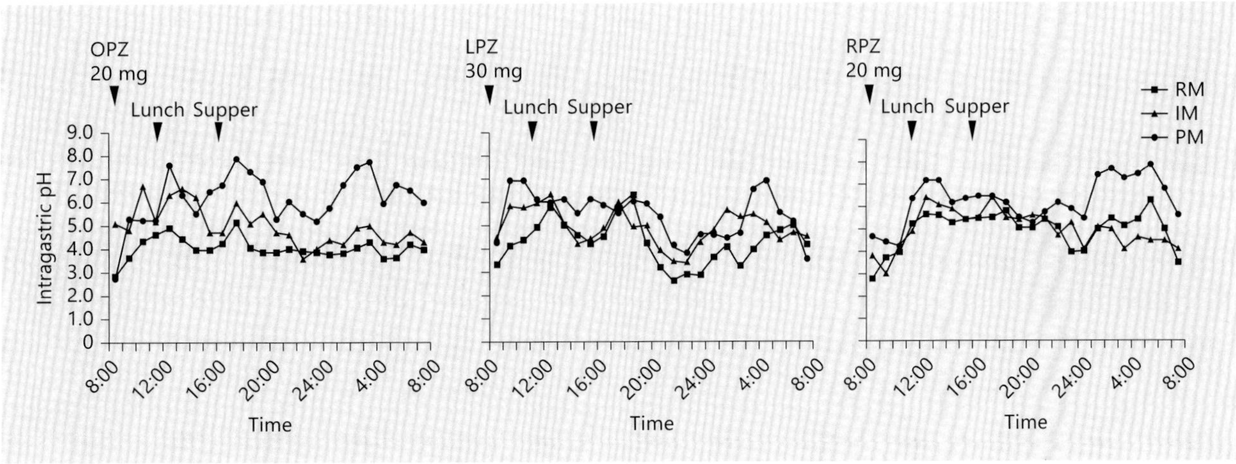

Fig. 4. Mean intragastric pH values as a function of the CYP2C19 genotype status for 20 mg of omeprazole (OPZ), 30 mg of lansoprazole (LPZ), or 20 mg of rabeprazole (RPZ) dosing for 8 days.

Table 3. Influence of CYP2C19 polymorphism on the cure rates of erosive esophagitis by different PPIs

PPI	RM		IM		PM		Statistics	Ref.
	%	n	%	n	%	n		
OPZ 20 mg for 4 weeks	43	25	95	33			0.04	[18]
LPZ 30 mg for 8 weeks	77.4	31	95.0	40	100.0	17	<0.05	[19]
LPZ 30 mg for 8 weeks	45.8	24	67.9	28	84.6	13	<0.05	[20]
EPZ 40 mg for 4 weeks	50.7	148	43.1	51	50.0	6	NS	[24]
RPZ 10 mg for 8 weeks	86.1	36	92.0	50	82.4	17	NS	[25]

OPZ = Omeprazole; LPZ = lansoprazole; RPZ = rabeprazole.

Table 4. Impact of different PPIs on the influence of CYP2C19 polymorphism on the eradication rates of *H. pylori* by triple therapy with a PPI, amoxicillin, and clarithromycin

PPI	RM	IM	PM	p	Ref.
OPZ	60.3	84.0	100.0	0.046	[28]
OPZ and LPZ	72.7	92.1	97.8	<0.001	[29]
LPZ	63	87	100	0.014	[30]
LPZ	58.5	67.3	76.2	<0.05	[31]
LPZ	57.8	88.2	92.3	<0.001	[32]
RPZ	85.9	88.7	96.3	0.055	[33]
RPZ	90.6	69.6	76.9	0.087	[34]
EPZ	92.9	93.3	95.0	>0.2	[35]

Values represent percentages. OPZ = Omeprazole; LPZ = lansoprazole; RPZ = rabeprazole.

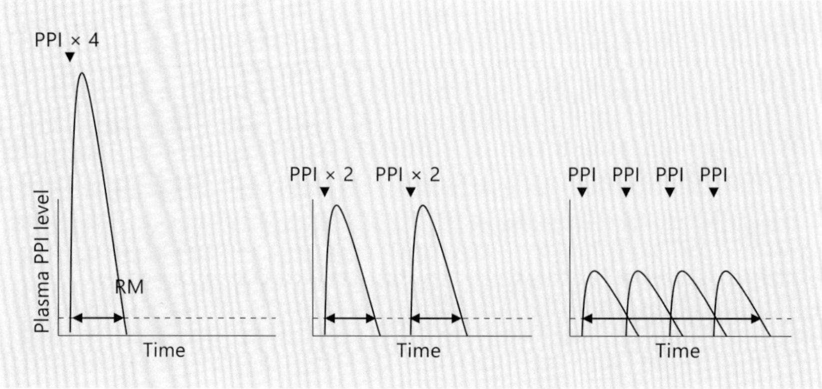

Fig. 5. Schematic model of plasma concentration of PPIs by different dosing schedules. When a PPI is dosed frequently (e.g. 4 times daily), sufficient plasma levels of PPI are expected to be sustained for a longer time.

attain the sufficient acid inhibition and not to be influenced by *CYP2C19* genotypes [16]. Therefore, divided or frequent dosing of a PPI seems to minimize the influence of *CYP2C19* genotypic differences and attain sufficient acid inhibition even in RMs of CYP2C19. As a matter of fact, dual therapies with a PPI and amoxicillin dosed 4 times daily were reported to attain sufficient eradication rates without a second antimicrobial agent such as clarithromycin or metronidazole even in CYP2C19 RMs [37, 38].

There have been conflicting data on the influence of *CYP2C19**17 on the pharmacokinetics of PPIs [39–43]. The *CYP2C19**17 allele was first reported to be associated with the ultrarapid metabolism of a PPI from a Swedish group [5, 39]. However, more recent reports from different institutions have indicated that *CYP2C19**17-associated differences in pharmacokinetics of PPIs are not significant [40, 41]. A meta-analysis [44] of the impacts of *CYP2C19**17 on the clinical outcomes of PPIs has indicated that the clinical impact of *CYP2C19**17/*17 is almost the same as that of *CYP2C19**1/*1, although *CYP2C19**17 is associated with an increased risk of bleeding in patients treated with clopidogrel [7, 45]. Therefore, in contrast to data of clopidogrel, which is one of the representative substrates of CYP2C19, recent reports have indicated that assignment of *CYP2C19**17 homozygotes as UMs rather than RMs or extensive metabolizers is inadequate in the clinical setting of PPIs. Further accumulation of data is needed on the influence of the *CYP2C19**17 allele on the clinical outcomes of acid-related disorders treated with PPIs.

Accordingly, CYP2C19 genotype status is considered as one of the predictable determinants for the results of a PPI-based therapy for GERD and *H. pylori* infection. If CYP2C19 genotype status is determined before treatment, an optimal dose of a PPI can be prescribed, based on this predetermined pharmacogenomic status. Therefore, the individualized therapeutic strategy based on the individual CYP2C19 genotype status should be conducted in such a therapeutic manner as performed with an increased dose of a PPI, particularly in the patients with the RM genotype of CYP2C19. This predetermined therapeutic strategy should be expected to increase the cure rates achieved by the initial treatment [46].

Drug-Drug Interaction between Proton Pump Inhibitors and Other Drugs in Relation to CYP2C19 Polymorphisms

Effects of Proton Pump Inhibitors on the Activity of P450

PPIs are often coadministered with other drugs, including substrates of P450. The drug-drug interactions of PPIs with other drugs are of clinical importance. Stedman and Barclay [47] summarized the drug-drug interaction of different PPIs via cytochrome P450 and reported that there was some variation in their potential for drug interactions due to differences in enzyme inhibition. Kodaira et al. [48] reported the effect of different PPIs on the activity of P450 assessed by the [^{13}C]-aminopyrine breath test in relation to CYP2C19 polymorphisms. They reported that different PPIs inhibited cytochrome P450 activity to different extents.

Interaction between Proton Pump Inhibitors and Clarithromycin

One of the current regimens for eradication of *H. pylori* is the triple therapy with a PPI, amoxicillin, and clarithromycin [49]. Clarithromycin is not only metabolized by CYP3A4, but is also a potent inhibitor of CYP3A4. CYP3A4 is involved in the sulfoxidation of PPIs [50]. Clarithromycin also affects the activity of CYP2C19 [51]. Therefore, when a PPI and clarithromycin are coadministered, a drug-drug interaction between the PPI and clarithromycin can occur.

Figure 6 shows the effects of clarithromycin on plasma omeprazole levels as a function of CYP2C19 genotype status. Plasma omeprazole levels increase by coadministration with clarithromycin in each of the different CYP2C19 genotype groups. In particular, plasma omeprazole levels in PMs are extremely high because PMs lack CYP2C19; therefore, CYP3A4, which is an important PPI-metabolizing enzyme in PMs [52], is inhibited by clarithromycin. Another paper also demonstrates that clarithromycin increases plasma lansoprazole levels by inhibiting the CYP3A4 activity in patients who undergo triple therapy with lansoprazole, amoxicillin, and clarithromycin [53]. These results indicate that the drug-drug interaction between clarithromycin and a PPI may underlie the high cure rate for the eradication of *H. pylori* obtained by triple therapy with a PPI, amoxicillin, and clarithromycin. Plasma clarithromycin levels also differ among the different CYP2C19 genotype groups [52], which is assumed to be related to different eradication rates among the different CYP2C19 genotype groups.

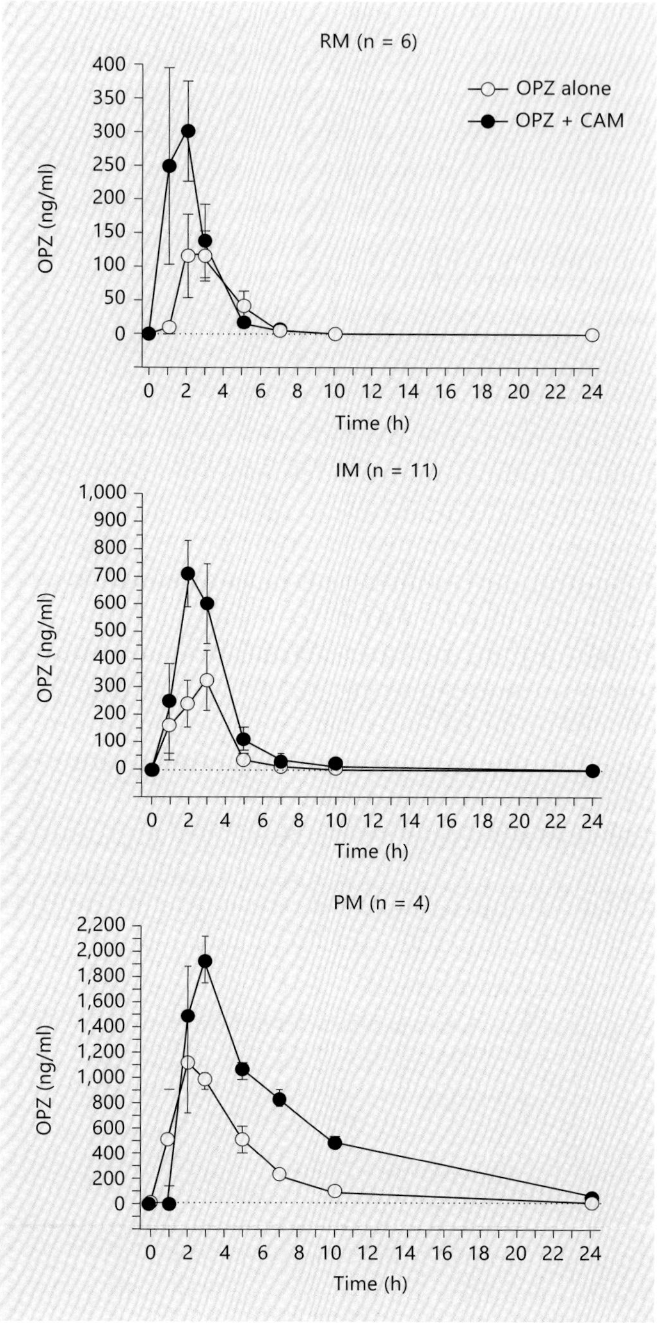

Fig. 6. Mean (±SE) plasma levels of omeprazole (OPZ) in the RM, IM, and PM groups with and without clarithromycin (CAM). Plasma concentrations of omeprazole were increased by CAM in the three different CYP2C19 genotype groups.

Interaction between Proton Pump Inhibitors and Clopidogrel

In 2008, the consensus and guidelines of the American College of Cardiology Foundation (ACCF), the American College of Gastroenterology (ACG), and American Heart Association (AHA) on antiplatelet therapy was published [54–56], and stated that patients with a risk of peptic ulcer and/or treated with two or more antiplatelet agents are recommended to be treated with a PPI. Therefore, patients treated with clopidogrel have a greater chance to be treated with a PPI. However, clopidogrel is activated with the metabolization by CYP2C19. Therefore, the plasma level of the active metabolite of clopidogrel depends on the activity of CYP2C19. In fact, the inhibitory effect of clopidogrel on platelet aggregation depends on CYP2C19 genotype status [57] and is also associated the clinical events [58]. Moreover, concomitant use of clopidogrel and a PPI induces a drug-drug interaction via CYP2C19, resulting in decreased activation of clopidogrel. Juurlink et al. [59] reported that concomitant therapy with a PPI was associated with an increased risk of reinfarction. Ho et al. [60] reported that concomitant use of clopidogrel and a PPI was associated with an increased risk of adverse outcomes compared to the use of clopidogrel without a PPI, and they suggested that the use of a PPI might be associated with attenuation of benefits of clopidogrel. Together, there are merits (i.e. gastric protection) and demerits (i.e. attenuation of clopidogrel efficacy) in the concomitant use of a PPI in patients undergoing antiplatelet therapy including clopidogrel. This is the therapeutic dilemma of prophylaxis use of PPI in antiplatelet therapy.

Furuta et al. [61] studied the effects of omeprazole, lansoprazole, and rabeprazole on the antiplatelet functions of clopidogrel in relation to CYP2C19 genotype status and found that the three PPIs attenuated the antiplatelet functions of clopidogrel in different degrees and that the levels of attenuation of clopidogrel by PPIs depended on CYP2C19 genotype status. Although the efficacy of clopidogrel was decreased by a PPI in RMs of CYP2C19, the levels of antiplatelet function of clopidogrel after attenuation by a PPI in this group were not mostly problematic [i.e. rarely decreased to the levels of 'low-responder' (inhibition of platelet aggregation <30%)]. On the other hand, conversion from responder to low responder easily occurs in patients with the IM genotype of CYP2C19. The subjects with the PM genotype of CYP2C19 were low responders irrespective of PPI use. Therefore, the influence of a PPI on the activity of clopidogrel also depends on the *CYP2C19* genotypes. When omeprazole and clopidogrel are dosed separately, the interaction cannot be avoided (fig. 6). However, in the RMs of CYP2C19, the interaction appeared small by separate dosing (fig. 7).

Although prospective clinical studies have indicated that the influence of PPIs on the clinical effect of clopidogrel is not problematic [62, 63], all pharmacokinetic studies have indicated the presence of a drug-drug interaction between PPIs and clopidogrel. Patients receiving dual antiplatelet therapy who are at risk of gastrointestinal bleeding need treatment with a PPI. However, clinicians must take this interaction into consideration for the careful follow-up of the patients.

Changes in gastric pH levels by a PPI sometimes influence the absorption of other drugs. Inhibition of gastric acid secretion increases the absorption of acid-labile drugs such as digoxin and sulfonylurea and weak acids such as diazepam, aspirin, and the diuretic furosemide. On the other hand, inhibition of acid secretion decreases the ab-

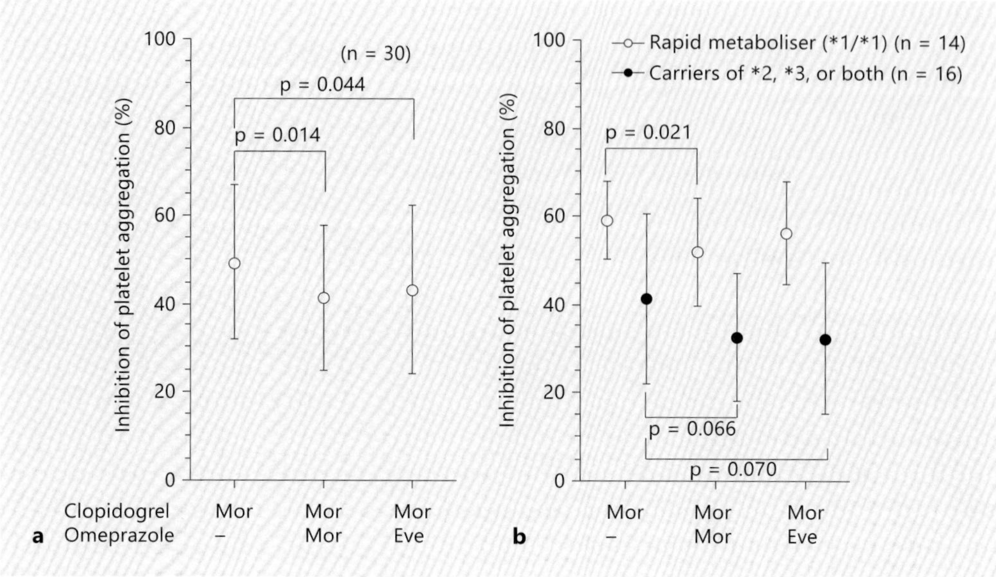

Fig. 7. a, b The effect of omeprazole (20 mg) dosed concomitantly in the morning (Mor) or separately in the evening (Eve) on the inhibition of platelet aggregation by clopidogrel (75 mg). Omeprazole dosed concomitantly attenuates the antiplatelet function of clopidogrel. This interaction cannot be avoided by separate dosing of omeprazole. **a** When data are stratified based on CYP2C19 status (carrier or noncarrier of the *2 and/or *3 allele of CYP2C19, it seems the interaction can be decreased to some extent in RMs of CYP2C19.

sorption of weak bases, such as ketoconazole, itraconazole, dipyridamole, tetracycline, cephalosporin antibiotics, gefitinib, and erlotinib [64–66]. To avoid this interaction, taking medicine with cola is recommended for ketoconazole because the pH of cola is around 2.5 and taking ketoconazole with cola improves the absorption of ketoconazole inhibited by omeprazole [67].

Conclusion

Not only pharmacokinetics and pharmacodynamics, but also clinical outcomes of PPI-based therapy are affected by the CYP2C19 genotype or pharmacogenomics. PPIs also influence the kinetics of other drugs by P450 interaction and/or acid inhibition. PPIs are now the first choice for acid inhibition. Moreover, in GERD patients, PPIs are dosed for a long time. The clinical effects of PPIs depend on the CYP2C19 genotype status of the patient as discussed in this review. Therefore, this genotyping test is a useful clinical tool for optimal PPI treatment. The cost-effectiveness of this genotyping test needs to be tested in a future study.

References

1 Furuta T, Shirai N, Sugimoto M, et al: Pharmacogenomics of proton pump inhibitors. Pharmacogenomics 2004;5:181–202.

2 De Morais SM, Wilkinson GR, Blaisdell J, et al: Identification of a new genetic defect responsible for the polymorphism of (S)-mephenytoin metabolism in Japanese. Mol Pharmacol 1994;46:594–598.

3 de Morais SM, Goldstein JA, Xie HG, et al: Genetic analysis of the S-mephenytoin polymorphism in a Chinese population. Clin Pharmacol Ther 1995;58:404–411.

4 de Morais SM, Wilkinson GR, Blaisdell J, et al: The major genetic defect responsible for the polymorphism of S-mephenytoin metabolism in humans. J Biol Chem 1994;269:15419–15422.

5 Sim SC, Risinger C, Dahl ML, et al: A common novel CYP2C19 gene variant causes ultrarapid drug metabolism relevant for the drug response to proton pump inhibitors and antidepressants. Clin Pharmacol Ther 2006;79:103–113.

6 Xie HG, Stein CM, Kim RB, et al: Allelic, genotypic and phenotypic distributions of S-mephenytoin 4'-hydroxylase (CYP2C19) in healthy Caucasian populations of European descent throughout the world. Pharmacogenetics 1999;9:539–549.

7 Strom CM, Goos D, Crossley B, et al: Testing for variants in CYP2C19: population frequencies and testing experience in a clinical laboratory. Genet Med 2012;14:95–100.

8 Rudberg I, Mohebi B, Hermann M, et al: Impact of the ultrarapid CYP2C19*17 allele on serum concentration of escitalopram in psychiatric patients. Clin Pharmacol Ther 2008;83:322–327.

9 Sugimoto K, Uno T, Yamazaki H, et al: Limited frequency of the CYP2C19*17 allele and its minor role in a Japanese population. Br J Clin Pharmacol 2008;65:437–439.

10 Kubota T, Chiba K, Ishizaki T: Genotyping of S-mephenytoin 4'-hydroxylation in an extended Japanese population. Clin Pharmacol Ther 1996;60:661–666.

11 Xie HG, Kim RB, Stein CM, et al: Genetic polymorphism of (S)-mephenytoin 4'-hydroxylation in populations of African descent. Br J Clin Pharmacol 1999;48:402–408.

12 Furuta T, Ohashi K, Kosuge K, et al: CYP2C19 genotype status and effect of omeprazole on intragastric pH in humans. Clin Pharmacol Ther 1999;65:552–561.

13 Shirai N, Furuta T, Moriyama Y, et al: Effects of CYP2C19 genotypic differences in the metabolism of omeprazole and rabeprazole on intragastric pH. Aliment Pharmacol Ther 2001;15:1929–1937.

14 Furuta T, Shirai N, Xiao F, et al: Effect of high-dose lansoprazole on intragastic pH in subjects who are homozygous extensive metabolizers of cytochrome P4502C19. Clin Pharmacol Ther 2001;70:484–492.

15 Nexium Capsule. Interview Form 2011; NEX8IF102 http://www1.astrazeneca-us.com/pi/Nexium.pdf.

16 Lou HY, Chang CC, Sheu MT, et al: Optimal dose regimens of esomeprazole for gastric acid suppression with minimal influence of the CYP2C19 polymorphism. Eur J Clin Pharmacol 2009;65:55–64.

17 Saitoh T, Watanabe Y, Kubo Y, et al: Effect of H2 blockers on the circadian rhythm of intragastric acidity. Biomed Pharmacother 2002;56(suppl 2):349s–352s.

18 Zendehdel N, Biramijamal F, Hossein-Nezhad A, et al: Role of cytochrome P450 2C19 genetic polymorphisms in the therapeutic efficacy of omeprazole in Iranian patients with erosive reflux esophagitis. Arch Iran Med 2010;13:406–412.

19 Kawamura M, Ohara S, Koike T, et al: The effects of lansoprazole on erosive reflux oesophagitis are influenced by CYP2C19 polymorphism. Aliment Pharmacol Ther 2003;17:965–973.

20 Furuta T, Shirai N, Watanabe F, et al: Effect of cytochrome P4502C19 genotypic differences on cure rates for gastroesophageal reflux disease by lansoprazole. Clin Pharmacol Ther 2002;72:453–460.

21 Kawamura M, Ohara S, Koike T, et al: Cytochrome P450 2C19 polymorphism influences the preventive effect of lansoprazole on the recurrence of erosive reflux esophagitis. J Gastroenterol Hepatol 2007;22:222–226.

22 Furuta T, Sugimoto M, Kodaira C, et al: CYP2C19 genotype is associated with symptomatic recurrence of GERD during maintenance therapy with low-dose lansoprazole. Eur J Clin Pharmacol 2009;65:693–698.

23 Saitoh T, Otsuka H, Kawasaki T, et al: Influences of CYP2C19 polymorphism on recurrence of reflux esophagitis during proton pump inhibitor maintenance therapy. Hepatogastroenterology 2009;56:703–706.

24 Schwab M, Klotz U, Hofmann U, et al: Esomeprazole-induced healing of gastroesophageal reflux disease is unrelated to the genotype of CYP2C19: evidence from clinical and pharmacokinetic data. Clin Pharmacol Ther 2005;78:627–634.

25 Ariizumi K, Ohara S, Koike T, et al: Therapeutic effects of 10 mg/day rabeprazole administration on reflux esophagitis was not influenced by the CYP2C19 polymorphism. J Gastroenterol Hepatol 2006;21:1428–1434.

26 Furuta T, Sugimoto M, Shirai N, et al: CYP2C19 pharmacogenomics associated with therapy of *Helicobacter pylori* infection and gastro-esophageal reflux diseases with a proton pump inhibitor. Pharmacogenomics 2007;8:1199–1210.

27 Zhao F, Wang J, Yang Y, et al: Effect of CYP2C19 genetic polymorphisms on the efficacy of proton pump inhibitor-based triple therapy for *Helicobacter pylori* eradication: a meta-analysis. Helicobacter 2008;13: 532–541.
28 Sapone A, Vaira D, Trespidi S, et al: The clinical role of cytochrome p450 genotypes in *Helicobacter pylori* management. Am J Gastroenterol 2003;98:1010–1015.
29 Furuta T, Shirai N, Takashima M, et al: Effect of genotypic differences in CYP2C19 on cure rates for *Helicobacter pylori* infection by triple therapy with a proton pump inhibitor, amoxicillin, and clarithromycin. Clin Pharmacol Ther 2001;69:158–168.
30 Okudaira K, Furuta T, Shirai N, et al: Concomitant dosing of famotidine with a triple therapy increases the cure rates of *Helicobacter pylori* infections in patients with the homozygous extensive metabolizer genotype of CYP2C19. Aliment Pharmacol Ther 2005;21:491–497.
31 Suzuki T, Matsuo K, Sawaki A, et al: Influence of smoking and CYP2C19 genotypes on *H. pylori* eradication success. Epidemiol Infect 2007;135:171–176.
32 Furuta T, Shirai N, Sugimoto M, et al: Influence of CYP2C19 pharmacogenetic polymorphism on proton pump inhibitor-based therapies. Drug Metab Pharmacokinet 2005;20:153–167.
33 Kuwayama H, Asaka K, Sugiyama T, et al: Rabeprazole-based eradication therapy for *Helicobacter pylori*: a large-scale study in Japan. Aliment Pharmacol Ther 2007;25:1105–1113.
34 Miki I, Aoyama N, Sakai T, et al: Impact of clarithromycin resistance and CYP2C19 genetic polymorphism on treatment efficacy of *Helicobacter pylori* infection with lansoprazole- or rabeprazole-based triple therapy in Japan. Eur J Gastroenterol Hepatol 2003;15:27–33.
35 Sheu BS, Kao AW, Cheng HC, et al: Esomeprazole 40 mg twice daily in triple therapy and the efficacy of *Helicobacter pylori* eradication related to CYP2C19 metabolism. Aliment Pharmacol Ther 2005;21:283–288.
36 Sugimoto M, Furuta T, Shirai N, et al: Different dosage regimens of rabeprazole for nocturnal gastric acid inhibition in relation to cytochrome P450 2C19 genotype status. Clin Pharmacol Ther 2004;76:290–301.
37 Furuta T, Sugimoto M, Kodaira C, et al: The dual therapy with 4 times daily dosing of rabeprazole and amoxicillin as the 3rd rescue regimen for eradication of *H. pylori*. Hepatogastroenterology 2010;57:1314–1319.
38 Miehlke S, Hansky K, Schneider-Brachert W, et al: Randomized trial of rifabutin-based triple therapy and high-dose dual therapy for rescue treatment of *Helicobacter pylori* resistant to both metronidazole and clarithromycin. Aliment Pharmacol Ther 2006; 24:395–403.
39 Baldwin RM, Ohlsson S, Pedersen RS, et al: Increased omeprazole metabolism in carriers of the CYP2C19*17 allele; a pharmacokinetic study in healthy volunteers. Br J Clin Pharmacol 2008;65:767–774.
40 Ohlsson Rosenborg S, Mwinyi J, Andersson M, et al: Kinetics of omeprazole and escitalopram in relation to the CYP2C19*17 allele in healthy subjects. Eur J Clin Pharmacol 2008;64:1175–1179.
41 Kearns GL, Leeder JS, Gaedigk A: Impact of the CYP2C19*17 allele on the pharmacokinetics of omeprazole and pantoprazole in children: evidence for a differential effect. Drug Metab Dispos 2010;38:894–897.
42 Gumus E, Karaca O, Babaoglu MO, et al: Evaluation of lansoprazole as a probe for assessing cytochrome P450 2C19 activity and genotype-phenotype correlation in childhood. Eur J Clin Pharmacol 2012;68: 629–636.
43 Kurzawski M, Gawronska-Szklarz B, Wrzesniewska J, et al: Effect of CYP2C19*17 gene variant on *Helicobacter pylori* eradication in peptic ulcer patients. Eur J Clin Pharmacol 2006;62:877–880.
44 Li-Wan-Po A, Girard T, Farndon P, et al: Pharmacogenetics of CYP2C19: functional and clinical implications of a new variant CYP2C19*17. Br J Clin Pharmacol 2010;69:222–230.
45 Sibbing D, Gebhard D, Koch W, et al: Isolated and interactive impact of common CYP2C19 genetic variants on the antiplatelet effect of chronic clopidogrel therapy. J Thromb Haemost 2010;8:1685–1693.
46 Furuta T, Shirai N, Kodaira M, et al: Pharmacogenomics-based tailored versus standard therapeutic regimen for eradication of *H. pylori*. Clin Pharmacol Ther 2007;81:521–528.
47 Stedman CA, Barclay ML: Review article: comparison of the pharmacokinetics, acid suppression and efficacy of proton pump inhibitors. Aliment Pharmacol Ther 2000;14:963–978.
48 Kodaira C, Uchida S, Yamade M, et al: Influence of different proton pump inhibitors on activity of cytochrome P450 assessed by [13C]-aminopyrine breath test. J Clin Pharmacol 2012;52:432–439.
49 Asaka M, Sugiyama T, Kato M, et al: A multicenter, double-blind study on triple therapy with lansoprazole, amoxicillin and clarithromycin for eradication of *Helicobacter pylori* in Japanese peptic ulcer patients. Helicobacter 2001;6:254–261.
50 Andersson T, Miners JO, Veronese ME, et al: Identification of human liver cytochrome P450 isoforms mediating secondary omeprazole metabolism. Br J Clin Pharmacol 1994;37:597–604.
51 Rodrigues AD, Roberts EM, Mulford DJ, et al: Oxidative metabolism of clarithromycin in the presence of human liver microsomes. Major role for the cytochrome P4503A (CYP3A) subfamily. Drug Metab Dispos 1997;25:623–630.

52 Furuta T, Ohashi K, Kobayashi K, et al: Effects of clarithromycin on the metabolism of omeprazole in relation to CYP2C19 genotype status in humans. Clin Pharmacol Ther 1999;66:265–274.
53 Ushiama H, Echizen H, Nachi S, et al: Dose-dependent inhibition of CYP3A activity by clarithromycin during *Helicobacter pylori* eradication therapy assessed by changes in plasma lansoprazole levels and partial cortisol clearance to 6beta-hydroxycortisol. Clin Pharmacol Ther 2002;72:33–43.
54 Bhatt DL, Scheiman J, Abraham NS, et al: ACCF/ACG/AHA 2008 expert consensus document on reducing the gastrointestinal risks of antiplatelet therapy and NSAID use: a report of the American College of Cardiology Foundation Task Force on Clinical Expert Consensus Documents. J Am Coll Cardiol 2008;52:1502–1517.
55 Bhatt DL, Scheiman J, Abraham NS, et al: ACCF/ACG/AHA 2008 expert consensus document on reducing the gastrointestinal risks of antiplatelet therapy and NSAID use. Am J Gastroenterol 2008;103:2890–2907.
56 Bhatt DL, Scheiman J, Abraham NS, et al: ACCF/ACG/AHA 2008 expert consensus document on reducing the gastrointestinal risks of antiplatelet therapy and NSAID use: a report of the American College of Cardiology Foundation Task Force on Clinical Expert Consensus Documents. Circulation 2008;118:1894–1909.
57 Umemura K, Furuta T, Kondo K: The common gene variants of CYP2C19 affect pharmacokinetics and pharmacodynamics in an active metabolite of clopidogrel in healthy subjects. J Thromb Haemost 2008;6:1439–1441.
58 Collet JP, Hulot JS, Pena A, et al: Cytochrome P450 2C19 polymorphism in young patients treated with clopidogrel after myocardial infarction: a cohort study. Lancet 2009;373:309–317.
59 Juurlink DN, Gomes T, Ko DT, et al: A population-based study of the drug interaction between proton pump inhibitors and clopidogrel. CMAJ 2009;180:713–718.
60 Ho PM, Maddox TM, Wang L, et al: Risk of adverse outcomes associated with concomitant use of clopidogrel and proton pump inhibitors following acute coronary syndrome. JAMA 2009;301:937–944.
61 Furuta T, Iwaki T, Umemura K: Influences of different proton pump inhibitors on the anti-platelet function of clopidogrel in relation to CYP2C19 genotypes. Br J Clin Pharmacol 2010;70:383–392.
62 O'Donoghue ML, Braunwald E, Antman EM, et al: Pharmacodynamic effect and clinical efficacy of clopidogrel and prasugrel with or without a proton-pump inhibitor: an analysis of two randomised trials. Lancet 2009;374:989–997.
63 Bhatt DL, Cryer BL, Contant CF, et al: Clopidogrel with or without omeprazole in coronary artery disease. N Engl J Med 2010;363:1909–1917.
64 Reynolds JC: The clinical importance of drug interactions with antiulcer therapy. J Clin Gastroenterol 1990;12(suppl 2):S54–S63.
65 Humphries TJ, Merritt GJ: Review article: drug interactions with agents used to treat acid-related diseases. Aliment Pharmacol Ther 1999;13(suppl 3):18–26.
66 Jaruratanasirikul S, Sriwiriyajan S: Effect of omeprazole on the pharmacokinetics of itraconazole. Eur J Clin Pharmacol 1998;54:159–161.
67 Chin TW, Loeb M, Fong IW: Effects of an acidic beverage (Coca-Cola) on absorption of ketoconazole. Antimicrob Agents Chemother 1995;39:1671–1675.

Takahisa Furuta, MD, PhD
Center for Clinical Research, Hamamatsu University School of Medicine
1-20-1 Handa-Yama
Hamamatsu, 431-3192 (Japan)
E-Mail tafuruta@fd6.so-net.ne.jp

Chiba T, Malfertheiner P, Satoh H (eds): Proton Pump Inhibitors: A Balanced View.
Front Gastrointest Res. Basel, Karger, 2013, vol 32, pp 34–46 (DOI: 10.1159/000350628)

Proton Pump Inhibitors in Gastroesophageal Reflux Disease

Stanislas Bruley des Varannes

Institut des Maladies de l'Appareil Digestif, Centre Hospitalier Universitaire de Nantes, Nantes, France

Abstract

During the last two decades the therapeutic superiority of proton pump inhibitors (PPIs) over other drugs has dramatically changed the management of gastroesophageal reflux disease (GERD). Their important clinical efficiency as compared with previous drugs, especially H_2-receptor blockers, associated with an excellent safety profile has contributed to the huge commercial success of this class of drugs. Short-term therapeutic strategies are now well defined for dosages and for durations of 2, 4, or 8 weeks, according to the severity of symptoms as well as the associated mucosal lesions. Likewise, well-defined, long-term approaches for GERD treatment (on-demand, intermittent, or continuous) allow the prevention of symptom and lesion recurrence for this chronic, relapsing disease. Despite the excellent global efficiency of PPIs, however, there are still some unmet needs in GERD, as illustrated by unsatisfactory improvement in 20–40% of patients. Particularly in patients with atypical manifestations, additional studies are needed to better identify subgroups which could benefit from PPI therapy. In addition, despite their safety profile, further studies are needed to better define the long-term potential side effects or drug-drug interferences.

The high prevalence of gastroesophageal reflux disease (GERD) is well established in Western countries, with figures between 10 and 30% cited. Although the clinical expression of GERD may be wide, a useful delineation of GERD has been provided by the Montreal definition indicating that GERD is a condition that develops when reflux of the stomach contents into the esophagus causes troublesome symptoms and/or complications (fig. 1) [1]. The prevalence of GERD appears to be increasing, possibly related to Western epidemiological changes such as excess body weight and obesity. The consequences of GERD have been documented, with important impacts on quality of life and alteration in work productivity [2]. Although the disease is

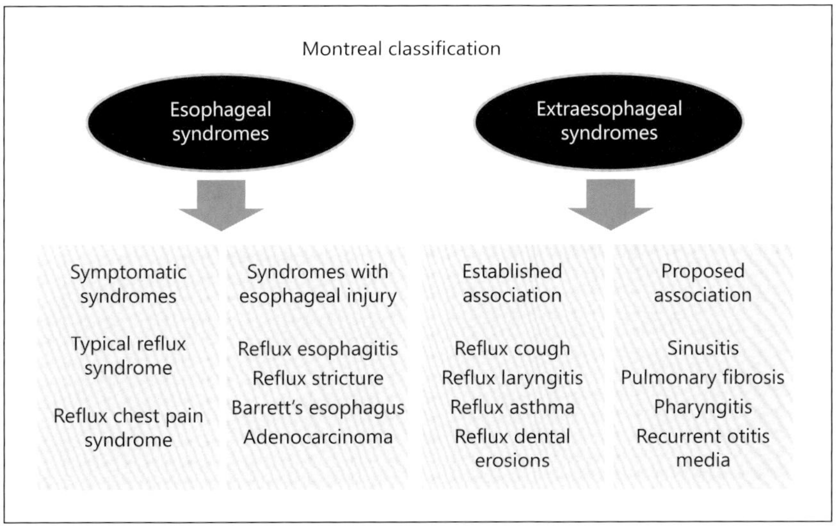

Fig. 1. Montreal classification of GERD syndromes developed by an international consensus group [1].

largely multifactorial in its pathophysiology, and primarily a motility disorder in which an impaired lower esophageal sphincter function plays a crucial role, the presence of acidic content in the esophagus is the main factor inducing symptoms and esophageal lesions. As a consequence, acid suppression has been widely demonstrated to be the current most effective drug-therapeutic approach fulfilling the main objectives of the treatment: alleviating symptoms, healing mucosal lesions, and restoring quality of life. In fact, with the availability of proton pump inhibitors (PPIs) in the 1990s, and due to their extensively documented pharmacological as well as clinical superiority over other drugs (antacids, prokinetics, and H_2-receptor antagonists), a dramatic change in the management of GERD has been observed [3, 4]. PPIs are now considered to be the mainstay of antireflux medical therapy, even if there are still some unmet needs in GERD therapy [5]. The objective of this chapter is to discuss the usual accepted strategies and results of PPI therapies in GERD using the main recommendations and largest clinical studies and meta-analyses in well-defined forms of the disease.

General Evaluation before Treatment of Gastroesophageal Reflux Disease

The cardinal symptoms of GERD, heartburn and regurgitation, are highly predictive for diagnosis of the disease. However, other widely prevalent symptoms may be associated with reflux of gastric content into the esophagus (fig. 1; Montreal classification). In clinical practice, when symptoms are considered to be possibly related to

GERD, the initial approach is a short medical management through acid control (either acid neutralization or acid suppression) [6]. Accordingly, in the majority of patients, when endoscopy is performed for any reason, this is more often after a period of treatment. White-light endoscopic examination allows the determination of the presence and severity of macroscopic esophageal injuries. Patients without evidence of mucosal injury are then classified as having nonerosive reflux disease (NERD), and patients with mucosal lesions are diagnosed with erosive esophagitis (EE) using the four-grade Los Angeles classification to characterize the degree of severity. In the near future, recent endoscopic technological advances such as narrow-band imaging or microconfocal endoscopy could be useful in helping to detect subtle mucosal changes and distinguish NERD patients who have microscopic lesions. Endoscopic biopsy samples may be performed either to determine the histological nature of an abnormal appearance or more systematically to look for abnormal tissue infiltration such as eosinophilic cells. Due to this frequently adopted clinical strategy, it is logical to consider separately the treatment of the two main situations defined according to the results of endoscopic examination, i.e. NERD and EE.

Proton Pump Inhibitors in the Treatment of Nonerosive Reflux Disease

NERD is the most frequently encountered situation, representing about two thirds of the patients and probably more when considering numerous minor clinical forms without endoscopic examination. Although lifestyle recommendations may be helpful, their efficiency is limited and they are difficult to follow in the long term. In fact, these patients are seen mainly in primary care and are often given an initial empirical acid-suppressive therapy. However, not all patients with NERD need antisecretory drugs. Recommendations are mainly based on the frequency and severity of symptoms, preferring a gradually increasing drug approach. Over-the-counter antacids/alginates represent the first level in patients with intermittent and occasional symptoms – i.e. once a week or less [7]. Then, although H_2-receptor antagonists may be given, PPIs often seem to be preferred now due to their overall higher efficiency [3]. In addition, at this stage physicians are tempted to consider and use this initial treatment as an empirical diagnostic test despite a rather low specificity. Consequently, PPIs are widely used in NERD patients and meta-analyses show that PPIs are a more effective treatment than H_2-receptor antagonists to control NERD symptoms [8]. However, and despite their largely documented pharmacological efficacy, the results of PPI treatments in patients with NERD are far from optimal. In fact, it is now established that if NERD rarely progresses to EE over time, it is more refractory to PPI treatment than EE. Several studies have shown that only one third of NERD patients can be fully cured with a PPI [9], although the improvement rate was reported in some studies to be between 50 and 70% or higher [10, 11]. A part of this lower efficiency may be related to a suboptimal GERD diagnosis, as suggested by a recent meta-anal-

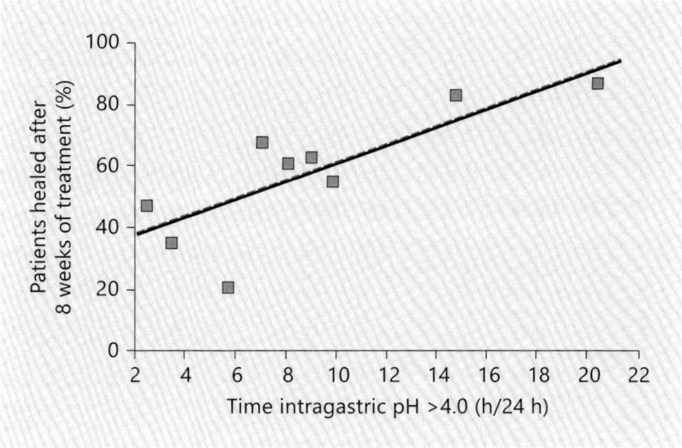

Fig. 2. Acid suppression therapy: the importance of acid control [17].

ysis showing that in well-defined GERD patients the estimated response rate after PPI therapy was similar to that observed in patients with EE [12].

The usual lower effectiveness of antisecretory drugs against symptoms in NERD relative to EE (approx. 20–30% lower) is related to different and multiple pathophysiologic factors in NERD (esophageal hypersensitivity, weakly acidic reflux, motility abnormalities, high proximal extent of reflux, psychological factors, etc.). In addition, some factors have been inconsistently reported as modulating the response to PPIs in NERD, such as the presence of other functional digestive disorders and a low BMI [13, 14] associated with a poorer response, whereas *Helicobacter pylori* infection and male gender have been reported to be associated with a better response [11].

Recently, and due to their excellent safety profile, PPIs have become available for over-the-counter use in many countries. According to the recommendations made by an expert group, a short course of over-the-counter PPIs (14 days) is suitable and efficacious in the short-term relief of mild and uncomplicated typical symptoms, and does not induce additional risk as compared with previously existing therapies (antacids, H_2-receptor antagonists) [15].

Proton Pump Inhibitors in the Treatment of Erosive Esophagitis

In EE, mucosal healing is the primary endpoint in clinical trials. The efficiency of PPIs in mucosal healing has been demonstrated extensively [4]. Overall, the rate of healing is related to the initial severity of lesions, reaching more than 90% in grade A-B and around 60–70% in grade C-D at 8 weeks [16]. Likewise, the lower the severity, the faster the speed of healing. pH-metric studies have also established a correlation between the level of esophageal acid exposure and the rate of healing (fig. 2) [17]. Overall, it is considered that therapeutic effectiveness of the different available PPIs is more or less equivalent, although greater intragastric suppression was observed in some studies for some PPIs. As PPIs are marketed at different licensed doses, direct comparison is not fully appropriate. Hence, it has been demonstrated in a very large study including more than 3,000 patients that 40 mg esomeprazole induced a higher healing rate than 30 mg lansoprazole and 20 mg omeprazole, mainly in patients with more severe grades of EE [18, 19].

In short-term PPI therapy, where the main objectives are healing of mucosal breaks and rapid symptom relief, the usual recommendations indicate the prescription of daily PPIs for 4–8 weeks based on the designs of clinical trials in minor or moderate forms of EE (Los Angeles grade A–B). More-prolonged administration (8 weeks) is recommended in severe forms (Los Angeles grade C–D) [4, 7]. If esophagitis does not heal and/or symptoms persist on standard-dose PPI therapy, it is common practice to double the dose of the PPI. However, even if this increase is associated with greater healing, the effect is considered to be modest [4].

In clinical practice, the absence of reflux symptoms after PPI therapy may be considered to be a good predictor of esophagitis healing [20]. However, in some studies the rate of control of symptoms is reported to be lower than that of mucosal healing. In fact, even in the presence of EE, and despite its high specificity confidently establishing the diagnosis of GERD, achieving complete symptom relief is more challenging than healing esophagitis. As in patients with NERD, several mechanisms may contribute to this lower efficacy against symptoms, such as esophageal hypersensitivity and/or mucosal permeability abnormalities, or the persistence of nonacid reflux or motility disorders.

In Barrett's esophagus, PPIs are used both for controlling symptoms and for healing the possibly associated esophagitis. Several studies have attempted to look for an effect of long-term PPI use on the extent of Barrett's metaplasia and on the risk of developing adenocarcinoma. No clear effect has been established regarding the regression of Barrett's esophagus. With respect to the course of dysplasia, and despite some results showing a reduction of dysplastic cell proliferation in experimental models, there is no direct evidence that long-term, high-dose PPI therapy is associated with a clear reduction in the risk of disease progression to high-grade dysplasia. However, PPI therapy may reduce and control low-grade dysplasia, justifying the prescription of PPI therapy (usually double dose) in this context before further histologic controls [21]. Finally, PPI-induced profound acid inhibition has also been associated with endoscopic ablative or resection therapy in patients with Barrett's esophagus.

Although patients with GERD have an increased risk of developing an esophageal stricture, with the increasing use of PPIs the occurrence of peptic esophageal strictures has become a rare event [22]. In addition, after endoscopic dilatation there is a reduced need for redilatation when PPI therapy is used, justifying the maintenance of long-term (perhaps lifelong) PPI therapy in patients with such a complication.

Proton Pump Inhibitors in Specific Conditions Related to Gastroesophageal Reflux Disease

Extraesophageal symptoms with a potential relation with GERD, especially laryngopharyngeal symptoms such as frequent throat clearing and hoarseness, chronic cough, and asthma (fig. 1), constitute a frequent reason for PPI prescription which most likely results in inappropriate and excessive PPI use, especially in primary care. Although noncardiac chest pain is grouped among the esophageal syndromes by the Montreal classification (fig. 1), it is not a symptom of typical GERD and consequently it is considered with extraesophageal symptoms. When extraesophageal symptoms are predominant, and in the absence of a direct and clear etiology for a direct organic disease, the potential relation with GERD is increasingly evoked. Logically, and due to the complex pathophysiology and multifactorial etiologies of these extraesophageal manifestations, patients are clearly less responsive to acid suppression than those with esophageal syndromes, especially when these symptoms are isolated (i.e. without heartburn or regurgitation).

As far as ENT symptoms are concerned, PPI treatment is the standard of care if GERD is suspected as the origin in patients with chronic throat symptoms. This strategy is reinforced by the consideration by otolaryngologists that frequent and nonspecific signs such as laryngeal edema and erythema are highly suggestive of GERD. However, a well-conducted, randomized, double-blind, placebo-controlled study failed to show a benefit of 40 mg esomeprazole b.i.d. for a duration of 16 weeks [23]. In fact, the majority of systematic reviews and meta-analyses have failed to establish the efficacy of twice-daily dosing of PPIs in these syndromes despite often prolonged treatment for several months. However, in these patients, and despite this insufficient evidence, the increasing incrimination of GERD as an etiologic factor has resulted in a widely used, pragmatic clinical strategy with twice-daily PPIs for 2 months. Secondarily, subjects with insufficient improvement (i.e. the majority) may be discussed for more-specific diagnostic tests if diagnostic suspicion remains high.

Regarding respiratory manifestations, although the role of GERD is established in some patients, the majority of therapeutic trials and systematic reviews have failed to demonstrate a significant effect of acid-suppressive treatments [24]. A recent randomized, double-blind, controlled clinical trial in a large series of patients with poorly controlled asthma who had minimal or no symptoms of GERD failed to show any effect of a twice-daily dose of 40 mg esomeprazole for 24 weeks on the occurrence of episodes of poor asthma control. Likewise, there was no effect on pulmonary function, airway reactivity, nocturnal awakening, or quality of life. Finally, the presence of asymptomatic gastroesophageal reflux documented by pH monitoring was not predictive of a treatment effect, suggesting that asymptomatic gastroesophageal reflux may not be a frequent cause of poor asthma control [25]. However, it is noteworthy that subgroup analysis and clinical practice suggest that PPIs may benefit some patients, although mainly when some reflux symptoms are present.

Contrary to other extraesophageal syndromes, GERD is the next most likely etiology in patients with unexplained chest pain once ischemic heart disease has been thoroughly eliminated. Several randomized, controlled trials have shown the benefit of twice-daily PPIs [26]. Current statements recommend the treatment of these patients using this approach and the performance of diagnostic testing with esophageal manometry and pH or impedance-pH monitoring, on treatment, in those who continue to experience chest pain despite this course of therapy [6].

In addition to the extraesophageal syndromes, the role of nocturnal reflux GERD in the pathogenesis of sleep disturbances has been increasingly recognized. In fact, the occurrence of nocturnal reflux symptoms has been established in at least two thirds of patients with GERD. As demonstrated using simultaneous polysomnographic and esophageal pH monitoring, these reflux episodes determine sleep arousals and overall alteration of sleep quality. PPIs have been shown not only to reduce nighttime heartburn, but also to improve sleep quality and work productivity [27].

Proton Pump Inhibitor Nonresponders: Refractory Gastroesophageal Reflux Disease

It can be estimated that 20–40% of reflux patients are dissatisfied with a standard PPI dose, reporting either partial or complete lack of response. The generally accepted definition of poor response to PPIs is an insufficient improvement (<50%) of the principal complaint after at least 8–12 weeks of PPI therapy. A growing practice recommended by the American Gastroenterological Association Guidelines for GERD is to double the PPI dose, assuming that more potent acid suppression will result in symptom resolution [6]. However, this strategy is frequently disappointing, with the majority of patients continuing to experience reflux symptoms. First, it has been established that, compared to patients with EE, patients with NERD more often fail PPI treatment, with about a 20% reduction of therapeutic gain [9]. Some other factors and patient characteristics have been found to be associated with a lower response to PPI in GERD, such as atypical GERD symptoms, the presence of other functional digestive disorders, BMI <25, disease duration, and *H. pylori* status [13, 28]. Likewise, some predominant symptoms may be associated with a lower response rate. Thus, patients with severe regurgitation have a lower rate of PPI response than patients with heartburn [29]. Finally patients with functional heartburn who have to be distinguished from NERD patients, even if they still represent a heterogeneous group with various underlying mechanisms, likely make up a large part of the PPI failures [30].

In recent years increasing interest has been given to patients who partially or completely fail to respond to PPI therapy [9, 31]. A rigorous and systematic approach is needed to determine the underlying mechanisms for this lack of response. To begin with, as misdiagnosis is likely to be common, it is of primary importance to establish the diagnosis of GERD with a high level of certainty. It is then usual to

Table 1. Main proposed underlying mechanisms and factors associated with a poor or absent response to PPIs in GERD patients

Misdiagnosis
Functional heartburn
Functional dyspepsia
Rumination syndrome – aerophagia
Achalasia
Infectious esophagitis
Eosinophilic esophagitis
Pill-induced esophagitis
Insufficient acid suppression
Lack of compliance to PPI therapy
Improper dosing time
Reduced oral bioavailability
Insufficient duration (NAB occurrence)
Genetic variations (rapid PPI metabolism)
True resistance to PPIs
Underlying pathophysiology
Weakly acidic reflux
Duodenogastroesophageal reflux
Delayed gastric emptying
Predominant regurgitation
Obesity
Comorbidities
Acid hypersecretion syndromes (ZES)
Concomitant FBDs
Psychological comorbidity

NAB = Nocturnal acid breakthrough; ZES = Zollinger-Elisson syndrome; FBDs = functional bowel disorders. Reproduced from [9].

consider that refractoriness to PPIs can be reflux-/nonreflux-related and patient- or therapy-related [32]. The main factors associated with a poor or absent response to PPIs in GERD patients are indicated in table 1 [9]. Strategies for diagnosing and managing patients with reflux symptoms refractory to PPIs are out of the scope of this chapter and can be found in several detailed reviews [31–33]. However, before embarking on other investigations, some empirical changes in PPI regime may represent a way of optimizing PPI treatment. In addition to checking for full compliance, it is of interest to repeat advice on when and how to take the medication to maximize the pharmacodynamic efficiency (i.e. 15–60 min before a meal). In fact, a recent US survey showed that only 27% of patients dosed their PPI correctly [34]. Likewise some adaptations, such as modifying the timetable of PPI intake (i.e. evening instead of morning), splitting the dose into two intakes (morning and evening), or even changing the formulation given, may help to improve the therapeutic response.

Being overweight/obese is associated with an increased risk of GERD and its complications, and may represent a negative prognostic factor with a higher rate of symptom relapse during maintenance therapy [35]. In addition to direct mechanical factors, some proinflammatory signals derived from the visceral adipose tissue are probably involved in the specific pathophysiology of these patients. However, in spite of this well-documented association, the impact of being overweight/obesity on the efficacy of PPIs has not been fully documented. Some studies have reported that patients with EE and increased BMI were significantly more likely to need continued acid-suppressive treatment or were more exposed to delayed healing of esophagitis [36]. In fact, to date no study has specifically investigated whether a high BMI affects the pharmacodynamic and pharmacokinetic profile of PPIs, and it remains inconclusive whether PPI doses should be adjusted according to body weight.

As refractory GERD still represents an important unmet clinical need, other alternatives are being developed to address PPI-resistant patients, such as new PPI formulations aimed at a faster onset of action and/or an extended acid-suppressive effect. Some of these new compounds such as potassium-competitive acid blockers, provided that their safety is established, might be helpful to better control nocturnal intragastric acidity. The use of an inhibitor of the occurrence of transient lower esophageal sphincter relaxations represented an attractive add-on therapy, but the disappointing clinical benefits in recent clinical trials led companies to stop their development.

Long-Term Gastroesophageal Reflux Disease Management with Proton Pump Inhibitors

Initial or regular PPI treatment is considered most often with the full marketed dosage (e.g. 20 mg omeprazole or rabeprazole, 30 mg lansoprazole, pantoprazole, or esomeprazole) for a few weeks (2 or 4 and usually <8 weeks) and then stopped. As PPIs do not correct the underlying pathophysiological abnormalities responsible for GERD, symptoms and mucosal healing are frequently chronically relapsing. Hence, a number of patients are unable to achieve acid suppression for more than a few of days or weeks and long-term acid suppressive therapy is needed. In fact, the occurrence and severity of relapses are highly variable between patients and necessitate adaptation of the strategy according to three main modalities: continuous, intermittent, or on-demand treatment (fig. 3). In clinical practice the determination of the best choice needs to best describe the natural course of symptoms in a given patient when stopping regular PPI treatment.

In patients with mild or moderate symptoms and infrequent relapses, intermittent short cycles (about 2 weeks) of daily PPI administration are usually suitable at each symptomatic relapse. In patients with frequent and close relapses, the choice is be-

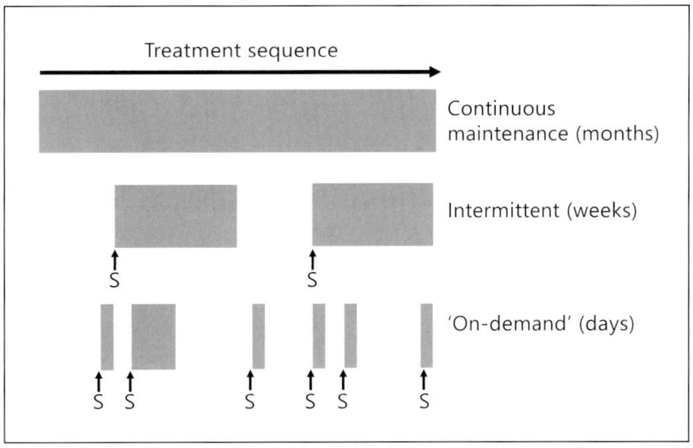

Fig. 3. Schematic representation of the three long-term strategies potentially available for the treatment of GERD. S = System recurrence.

tween a continuous daily treatment and a self-managed, on-demand strategy. In this last option, the patient adapts the PPI intake to the occurrence of symptoms. Several randomized controlled clinical trials have established the efficiency of the on-demand approach in the prevention of relapses [37]. In addition, on-demand treatment allows the reduction of PPI doses as patients are taking drugs on average 1 out of every 3 days, allowing a significant reduction in the cost of treatment. Overall, some data show that intermittent and on-demand strategies are suitable for 60% of newly diagnosed patients with GERD for the long-term management of symptomatic GERD of mild/moderate severity [38].

In other patients, more often patients with moderate/severe symptoms and moderate-to-severe esophagitis, GERD symptoms reappear systematically a few days after discontinuation of acid suppression. These patients are clearly PPI-dependent and need either a continuous PPI treatment or discussion of other therapeutic options such as antireflux surgery. Both strategies have been compared in the LOTUS randomized trial in patients initially responding to a PPI course with a 5-year follow-up [39]. There was a remission rate greater than 90% in the medical arm (85% in the surgical arm). However, long-term treatment has to be adapted to the natural course of the symptoms of each patient and regular fine-tunings will be required, especially in the initial determination of the best maintenance therapy strategy. Thus, regarding the dose, long-term PPI therapy is generally started with half-dosage, except for omeprazole, after an initial therapeutic sequence inducing control of symptoms and mucosal healing. This dosage, however, as well as the rhythm of PPI intake, has to be adapted according to the clinical response and the satisfaction of patient. In the LOTUS study, for example, increasing the daily dose was necessary in about one fourth of the patients to maintain the control of symptoms.

On the whole, PPI maintenance is now a safe and well-validated strategy demonstrating effectiveness in reducing the relapse rate and maintaining patients in remission with a nearly normal quality of life for very long time periods. In addition to these improvements, the beneficial effects of maintenance therapy in the prevention of complications are likely to be important as illustrated by the decreased prevalence of esophageal strictures in recent decades.

Conclusions

During the two last decades, the therapeutic superiority of PPIs over other drugs has dramatically changed the management of GERD. Their efficiency associated with an excellent safety profile has contributed to the huge commercial success of this class of drugs. Short-term therapeutic strategies are now well defined and adapted according to the severity of symptoms and the associated mucosal lesions. Likewise, long-term approaches for GERD treatment allow the prevention of symptoms and lesion recurrence in this chronic, relapsing disease in a large proportion of patients. However, despite the excellent efficiency of PPIs, there are still some unmet needs in GERD. Particularly in patients with atypical manifestations, additional studies are needed to better identify subgroups which could benefit from PPI therapy. In addition, despite their safety profile, further studies are needed to better define the long-term potential side effects or drug-drug interferences (see chapters of Nakamura et al. [pp. 92–101] and Kinoshita and Ishihara [pp. 102–115]).

References

1 Vakil N, Van Zanten SV, Kahrilas P, Dent J, Jones R, Global Consensus Group: The Montreal definition and classification of gastroesophageal reflux disease: a global evidence-based consensus. Am J Gastroenterol 2006;101:1900–1920.

2 Camilleri M, Dubois D, Coulie B, Jones M, Kahrilas PJ, Rentz AM, Sonnenberg A, Stanghellini V, Stewart WF, Tack J, Talley NJ, Whitehead W, Revicki DA: Prevalence and socioeconomic impact of upper gastrointestinal disorders in the United States: results of the US Upper Gastrointestinal Study. Clin Gastroenterol Hepatol 2005;3:543–552.

3 Chiba N, De Gara CJ, Wilkinson JM, Hunt RH: Speed of healing and symptom relief in grade II to IV gastroesophageal reflux disease: a meta-analysis. Gastroenterology 1997;112:1798–1810.

4 Khan M, Santana J, Donnellan C, Preston C, Moayyedi P: Medical treatments in the short term management of reflux oesophagitis. Cochrane Database Syst Rev 2007;2:CD003244.

5 Bruley des Varannes S, Coron E, Galmiche JP: Short and long-term PPI treatment for GERD. Do we need more-potent anti-secretory drugs? Best Pract Res Clin Gastroenterol 2010;24:905–921.

6 Kahrilas PJ, Shaheen NJ, Vaezi MF, Hitz SW, Black E, Modlin IM, Johnson SP, Allen J, Brill JV: American Gastroenterological Association medical position statement on the management of gastroesophageal reflux disease. Gastroenterology 2008;135:1383–1391.

7 AFSSAPS. Les antisécrétoires gastriques chez l'adulte. Recommandations de bonne pratique. 2007. http://www.snfge.asso.fr/01-bibliotheque/0d-pratiques-cliniques/reco_hors_has/antisecretoire-gastrique-argu.pdf.

8 van Pinxteren B, Sigterman KE, Bonis P, Lau J, Numans ME: Short-term treatment with proton pump inhibitors, H_2-receptor antagonists and prokinetics for gastro-oesophageal reflux disease-like symptoms and endoscopy negative reflux disease. Cochrane Database Syst Rev 2010;11:CD002095.

9 Scarpignato C: Poor effectiveness of proton pump inhibitors in non-erosive reflux disease: the truth in the end! Neurogastroenterol Motil 2012;24:697–704.

10 Armstrong D, Talley NJ, Lauritsen K, Moum B, Lind T, Tunturi-Hihnala H, Venables T, Green J, Bigard MA, Mössner J, Junghard O: The role of acid suppression in patients with endoscopy-negative reflux disease: the effect of treatment with esomeprazole or omeprazole. Aliment Pharmacol Ther 2004;20:413–421.

11 Hiyama T, Matsuo K, Urabe Y, Fukuhara T, Tanaka S, Yoshihara M, Haruma K, Chayama K: Meta-analysis used to identify factors associated with the effectiveness of proton pump inhibitors against non-erosive reflux disease. J Gastroenterol Hepatol 2009;24:1326–1332.
12 Weijenborg PW, Cremonini A, Smout AJ, Bredenood AJ: PPI therapy is equally effective in well-defined non-erosive reflux disease and in reflux esophagitis: a meta-analysis. Neurogastroenterol Motil 2012;24:747–757.
13 Zerbib F, Belhocine K, Simon M, Capdepont M, Mion F, Bruley des Varannes S, Galmiche JP: Clinical, but not oesophageal pH-impedance, profiles predict response to proton pump inhibitors in gastro-oesophageal reflux disease. Gut 2012;61:501–506.
14 Fletcher J, Derakhshan MH, Jones GR, Wirz AA, McColl KE: BMI is superior to symptoms in predicting response to proton pump inhibitor: randomised trial in patients with upper gastrointestinal symptoms and normal endoscopy. Gut 2011;60:442–448.
15 Haag S, Andrews JM, Katelaris PH, Gapasin J, Galmiche JP, Hunt R, Layer P, Malfertheiner P, Holtmann G: Management of reflux symptoms with over-the-counter proton pump inhibitors: issues and proposed guidelines. Digestion 2009;80:226–234.
16 Labenz J, Armstrong D, Lauritsen K, Katelaris P, Schmidt S, Schutze K, Wallner G, Juergens H, Preiksaitis H, Keeling N, Naucler E, Eklond S: A randomized comparative study of esomeprazole 40 mg versus pantoprazole 40 mg for healing erosive oesophagitis: the EXPO study. Aliment Pharmacol Ther 2005;21:739–746.
17 Bell NJ, Burget D, Howden CW, Wilkinson J, Hunt RH: Appropriate acid suppression for the management of gastro-oesophageal reflux disease. Digestion 1992;51(suppl 1):59–67.
18 Castell DO, Kahrilas PJ, Richter JE, Vakil NB, Johnson DA, Zuckerman S, et al: Esomeprazole (40 mg) compared with lansoprazole (30 mg) in the treatment of erosive esophagitis. Am J Gastroenterol 2002;97:575–583.
19 Richter JE, Kahrilas PJ, Johanson J, Maton P, Breiter JR, Hwang C, Marino V, Hamelin B, Levine JG: Efficacy and safety of esomeprazole compared with omeprazole in GERD patients with erosive esophagitis: a randomized controlled trial. Am J Gastroenterol 2001;96:656–665.
20 Sharma N, Donnellan C, Preston C, Delaney B, Duckett G, Moayyedi P: A systematic review of symptomatic outcomes used in oesophagitis drug therapy trials. Gut 2004;53:58–65.
21 El-Serag HB, Aguirre TV, Davis S, Kuebeler M, Bhattacharyya A, Sampliner RE: Proton pump inhibitors are associated with reduced incidence of dysplasia in Barrett's esophagus. Am J Gastroenterol 2004;99:1877–1883.
22 Ruigomez A, Garcia Rodriguez LA, Wallander MA, Johansson E, Eklund S: Esophageal stricture: incidence, treatment patterns, and recurrence rate. Am J Gastroenterol 2006;101:2685–2692.
23 Vaezi MF, Richter JE, Stasney CR, Spiegel JR, Iannuzzi RA, Crawley JA, Hwang C, Sostek MB, Shaker R: Treatment of chronic posterior laryngitis with esomeprazole. Laryngoscope 2006;116:254–260.
24 Galmiche JP, Zerbib F, Bruley des Varannes S: Review article: respiratory manifestations of gastro-oesophageal reflux disease. Aliment Pharmacol Ther 2008;27:449–464.
25 American Lung Association Asthma Clinical Research Centers, Mastronarde JG, Anthonisen NR, Castro M, Holbrook JT, Leone FT, Teague WG, Wise RA: Efficacy of esomeprazole for treatment of poorly controlled asthma. N Engl J Med 2009;360:1487–1499.
26 Dickman R, Emmons S, Cui H, Sewell J, Hernandez D, Esquivel RF, Fass R: The effect of a therapeutic trial of high-dose rabeprazole on symptom response of patients with non-cardiac chest pain: a randomized, double-blind, placebo-controlled, crossover trial. Aliment Pharmacol Ther 2005;22:547–555.
27 Johnson DA, Orr WC, Crawley JA, Traxler B, McCullough J, Brown KA, Roth T: Effect of esomeprazole on nighttime heartburn and sleep quality in patients with GERD: a randomized, placebo-controlled trial. Am J Gastroenterol 2005;100:1914–1922.
28 Dickman R, Boaz M, Aizic S, Beniashvili Z, Fass R, Niv Y: Comparison of clinical characteristics of patients with gastroesophageal reflux disease who failed proton pump inhibitor therapy versus those who fully responded. J Neurogastroenterol Motil 2011;17:387–394.
29 Kahrilas PJ, Howden CW, Hugues N: Response of regurgitation to proton pump inhibitor therapy in clinical trials of gastroesophageal reflux disease. Am J Gastroenterol 2011;106:1419–1425.
30 Surdea Blaga T, Dumitrascu D, Galmiche JP, Bruley des Varannes S: Functional heartburn: clinical characteristics and outcome. Eur J Gastroenterol Hepatol 2013;25:282–290.
31 Fass R, Sifrim D: Management of heartburn not responding to proton pump inhibitors. Gut 2009;58:295–309.
32 Dellon ES, Shaheen NJ: Persistent reflux symptoms in the proton pump inhibitor era: the changing face of gastroesophageal reflux disease. Gastroenterology 2010;139:7–13.

33 Sifrim D, Zerbib F: Diagnosis and management of patients with reflux symptoms refractory to proton pump inhibitors. Gut 2012;61:1340–1354.
34 Gunaratnam NT, Jessup TP, Inadomi J, Lascewski DP: Sub-optimal proton pump inhibitor dosing is prevalent in patients with poorly controlled gastro-oesophageal reflux disease. Aliment Pharmacol Ther 2006;23:1473–1477.
35 Labenz J, Armstrong D, Zetterstrand S, Eklund S, Leodolter A: Clinical trial: factors associated with freedom from relapse of heartburn inpatients with healed reflux oesophagitis – results from the maintenance phase of the EXPO study. Aliment Pharmacol Ther 2009;29:165–171.
36 Sheu BS, Chang WL, Cheng HC, Kao AW, Lu CC: Body mass index can determine the healing of reflux esophagitis with Los Angeles grades C and D by esomeprazole. Am J Gastroenterol 2008;103:2209–2214.
37 Pace F, Tonini M, Pallotta S, Molteni P, Porro GB: Systematic review: maintenance treatment of gastro-oesophageal reflux disease with proton pump inhibitors taken 'on-demand'. Aliment Pharmacol Ther 2007;26:195–204.
38 Bardhan DK: Intermittent and on-demand use of proton pump inhibitors in the management of symptomatic gastroesophageal reflux disease. Am J Gastroenterol 2003;9(suppl 3):S40–S48.
39 Galmiche JP, Hatlebakk J, Attwood S, Ell C, Fiocca R, Eklund S, Langström G, Lind T, Lundell L, LOTUS Trial Collaborators: Laparoscopic antireflux surgery vs esomeprazole treatment for chronic GERD: the LOTUS randomized clinical trial. JAMA 2011;305:1969–1977.

Stanislas Bruley des Varannes
IMAD – Service Gastroentérologie et Hépatologie – CHU Hôtel Dieu
FR–44093 Nantes Cedex (France)
E-Mail stanislas.bruleydesvarannes@chu-nantes.fr

Chiba T, Malfertheiner P, Satoh H (eds): Proton Pump Inhibitors: A Balanced View.
Front Gastrointest Res. Basel, Karger, 2013, vol 32, pp 47–58 (DOI: 10.1159/000350629)

Proton Pump Inhibitors in Gastroprotection: Prevention and Healing of Nonsteroidal Anti-Inflammatory Drug-, Aspirin-, and Cyclooxygenase-2 Inhibitor-Induced Gastroduodenal Lesions

Moe H. Kyaw · Francis K.L. Chan

Institute of Digestive Disease, The Chinese University of Hong Kong, Hong Kong, SAR, China

Abstract

Clinicians prescribing nonsteroidal anti-inflammatory drugs (NSAIDs) are challenged with two problems: (1) selection of nonselective traditional NSAIDs or cyclooxygenase-2 inhibitors, and (2) choice of suitable strategies to prevent the recurrence of drug-induced gastroduodenal injury and complications. This review will focus on the use of proton pump inhibitors (PPIs) in the prevention and healing of NSAID-induced gastroduodenal lesions. The role of PPIs will be compared with alternative gastroprotective agents and placebo.

Gastrointestinal (GI) adverse effects of nonsteroidal anti-inflammatory drugs (NSAIDs) including aspirin have been widely reported. The development of cyclooxygenase (COX)-2 inhibitors with the potential of minimizing GI toxic effects was welcomed by many clinicians. However, after reports of cardiovascular hazards related to COX-2 inhibitors, the trend in prescriptions of analgesia shifted back to the use of nonselective traditional NSAIDs (tNSAIDs) [1]. Subsequently, the extension of concerns of cardiovascular toxicity onto other classes of NSAIDS has raised further complications in the prescription of NSAIDs. Clinicians prescribing NSAIDs are challenged with two problems: (1) identifying patients at high-risk, and (2) the selection of suitable strategies to prevent recurrence of drug-induced gastroduodenal injury and complications. This review will focus on the use of proton pump inhibitors (PPIs) in the prevention and healing of NSAID-induced gastroduodenal lesions. For the analysis of the literature, endoscopic visualization of ulcers will be used as a surrogate endpoint for treatment with PPIs for the healing and prevention of NSAID-induced gastroduodenal lesions, with the exception of COX-2 inhibitors, where most studies only exist with ulcer bleeding as study endpoints. The treatment and prevention of GI bleeding using PPIs will be discussed in the chapter by Sung [pp. 68–76]. This review will evaluate each NSAID class (tNSAIDs, aspirin, COX-2 inhibitors), but an overlap of literature may exist as some studies have not differentiated the subclasses of NSAIDs.

Pathophysiology of Nonsteroidal Anti-Inflammatory Drug-Induced Peptic Ulcer Disease

NSAID-associated mucosal injury results from systemic prostaglandin inhibition associated with NSAID use causing a reduction in epithelial mucus secretion, mucosal blood flow, epithelial cell proliferation, bicarbonate secretion, and mucosal resistance to injury. The losses of the protective mechanisms allow gastric acid to come into close contact with epithelial surface cells [2]. COX catalyzes the production of prostaglandins from arachidonic acid. There are two isoforms of COX enzymes with different affinities to NSAIDs. COX-1 is present in most tissue at low levels, acting as a housekeeping enzyme, while COX-2 is unregulated by proinflammatory cytokines, causing inflammation and pain. The inhibition of prostaglandins from COX-2 in inflamed tissue and COX-1 in normal gastric tissue results in anti-inflammatory and gastroduodenal toxic effects [3].

Aspirin (ASA), like other NSAIDs, inhibits COX enzyme reducing the synthesis of prostaglandins, and resulting in its analgesic effect. At low doses (<325 mg daily), it predominantly inhibits COX-1, preventing the synthesis of platelet thromboxane A_2, which is required for platelet aggregation. Compared to other NSAIDs, COX-1 inhibition by aspirin is irreversible, and the effect on the platelet remains throughout the platelet lifespan (8–10 days) [4]. Although the mechanism involved in aspirin-induced GI lesions are not clearly understood, especially in the small and large intestine, it is known that aspirin induces toxic effects by local mucosal injury and systemic effects via reduction in prostaglandin synthesis. The rapid onset of mucosal injury by aspirin suggests that the mechanism of mucosal damage does not solely depend on prostaglandins [5].

Proton Pump Inhibitors and Healing of Traditional Nonsteroidal Anti-Inflammatory Drug-Induced Gastroduodenal Lesions

Several studies have compared PPI with H_2-receptor antagonists (H_2-RAs) for healing of tNSAID-induced gastroduodenal lesions in patients who required continuous treatment with tNSAIDs. They have all shown similar results, with PPIs being superior to H_2-RAs in healing tNSAID-induced gastroduodenal lesions. A large multicenter study (n = 541) compared two doses of omeprazole (20 and 40 mg/day) with ranitidine (40 mg/day), showing PPI to be more effective with gastric ulcer healing success of 84, 87, and 67%, respectively [6]. Similar results were produced with duodenal ulcer patients, with ulcer healing success of 92, 88, and 81%, respectively. There was no placebo in the study. This study also examined omeprazole as a preventive therapy for tNSAID-induced gastroduodenal lesions; the outcome of the study will be discussed later in the review. The superiority of PPIs over H_2-RAs was further supported in a prospective, double-blind, multicenter, parallel-group study (n = 353) [7]. At week 8, the gastric ulcer healing rate was 73% in patients treated with lansoprazole (30 mg/day), in comparison to a healing rate of 53% in those offered ranitidine (150 mg twice daily) ($p < 0.05$). Another large multicenter, randomized, double-blind study (n = 406) compared esomeprazole (20 and 40 mg/day) with ranitidine (150 mg twice daily) for the healing of gastric ulcers in patients who continued to receive daily tNSAID therapy [8]. At week 8, the gastric ulcer healing rates with esomeprazole 40 and 20 mg were 91.5 and 88.4%, respectively, significantly higher than 74.2% with ranitidine ($p < 0.01$). The gastric ulcer healing rates at 4 weeks (78.3 and 79.0% in the esomeprazole 20 and 40 mg/day groups, respectively) were also significantly higher ($p < 0.05$) than in the ranitidine group (66.7%). In a subsequent study from the same authors (n = 440), significant differences were already seen at 4 weeks into treatment with esomeprazole [9]. However, there was no significance difference in gastric ulcer healing rates from ranitidine (84.8, 85.7, 76.3%, in esomeprazole 20, 40 mg/day and ranitidine 150 mg twice daily, respectively) after 8 weeks of treatment with esomeprazole.

Concomitant treatment with PPIs for tNSAID-induced gastroduodenal lesions has also been compared with cytoprotective drugs such as sucralfate and misoprostol. In a small single-blind study (n = 88), omeprazole (20 mg/day) was shown to be superior to sucralfate in healing tNSAID-induced gastroduodenal ulcers in patients who continued to take tNSIADs. Omeprazole was significantly superior to sucralfate in inducing gastric ulcer healing after both 4 (87 vs. 52%, p = 0.007) and 8 weeks (100 vs. 82%, p = 0.04) [10]. A large randomized double-blind study (n = 935) compared the efficacy of omeprazole (20 and 40 mg/day) and misoprostol (200 μg twice daily) in ulcer healing associated with tNSAIDs. Healing rates among patients with duodenal ulcers were higher with either dose of omeprazole than with misoprostol, whereas healing rates among patients with erosions alone were higher with misoprostol (p = 0.004 and 0.14, respectively) [11].

Proton Pump Inhibitors and Prevention of tNSAID-Induced Gastroduodenal Lesions

Short-Term Treatment with Proton Pump Inhibitors (<3 Months of Nonsteroidal Anti-Inflammatory Drug Exposure)

Several studies support the benefit of short-term treatment with PPIs to prevent tNSAID-induced lesions. One short-term study over 3 weeks (n = 114) compared

Table 1. Randomized controlled trials with PPIs as preventive therapy of tNSAID-induced lesions; head-to-head comparison with H_2-RA, misoprostol, and placebo

Year/ref.	Intervention	Study size, n	Follow-up, weeks	Patients with GI lesions, %	p value
1998 [10]	omeprazole (20 mg o.d.)	114	3	54	
	placebo			65	n.s.
2002 [12]	omeprazole (20 mg o.d.)	660	5	0	
	placebo			5	<0.05
1998 [15]	omeprazole (20 mg o.d.)	169	24	3.6	
	placebo			16.5	0.004
1996 [16]	omeprazole (20 mg o.d.)	175	12	4.7	
	placebo			16.7	<0.001
2000 [17]	pantoprazole (40 mg o.d.)	94	12	18	
	placebo			45	0.036
2006[1] [18]	esomeprazole (40 mg o.d.)	844	32	4.7	<0.001
	esomeprazole (20 mg o.d.)			5.3	<0.001
	placebo			20.4	
2006[2] [18]	esomeprazole (40 mg o.d.)	585	32	4.4	0.007
	esomeprazole (20 mg o.d.)			5.2	0.018
	placebo			12.3	
1998 [6]	omeprazole (20 mg o.d.)	541	32	18	
	ranitidine (120 mg b.d.)			41	0.04
2005 [20]	lansoprazole (15 mg o.d.)	194	24	7.7	
	famotidine (20 mg b.d.)			15.4	n.s.
1998 [11]	omeprazole (20 mg o.d.)	935	24	49	<0.001
	misoprostol (400 µg o.d.)			62	0.001
	placebo			73	
2008 [21]	pantoprazole (20 mg o.d.)	515	32	5	
	misoprostol (400 µg o.d.)			14	<0.001
2002 [22]	lansoprazole (15 mg o.d.)	637	12	20	
	lansoprazole (30 mg o.d.)			18	
	misoprostol (400 µg b.d.)			7	
	placebo			49	n.s.

o.d. = Once daily; b.d. = twice daily; n.s. = nonsignificant; [1] VENUS. [2] PLUTO.

omeprazole (20 mg/day) with placebo in the prevention of gastroduodenal lesions induced by tNSAIDs. At final endoscopy, 46% of the omeprazole group and 35% of the placebo group had normal gastroduodenal mucosa (p = n.s.) [10]. In a large randomized, multicenter trial in *Helicobacter pylori*-positive patients (n = 660), the use of omeprazole (20 mg/day) for 5 weeks in those patients requiring tNSAID therapy (with no past or current peptic ulcers) was compared with a placebo group. The occurrence of peptic ulcers during the study period was 0% in the omeprazole group, compared with 5% in the placebo group [12].

A retrospective study (n = 676) examined the risk of peptic ulcer associated with acute and chronic NSAID therapy in elderly subjects, and the influence of antisecretory treatment on this risk. The use of NSAIDs, including aspirin, as well as antisecretory drugs (H_2-RAs and PPIs) was evaluated by a structured interview. The risk of peptic ulcer, adjusted for age, gender, *H. pylori* infection, and antisecretory drug use, was higher in acute users [gastric ulcer: odds ratio (OR) = 4.47; duodenal ulcer: OR = 2.39] than in chronic users (gastric ulcer: OR = 2.80; duodenal ulcer: OR = 1.68). PPI treatment was associated with a reduced risk of peptic ulcer in both acute (OR = 0.70) and chronic (OR = 0.32) NSAID/aspirin users. Treatment with PPIs resulted in an absolute risk reduction of peptic ulcer by 36.6% in acute and 34.6% in chronic NSAID users. The number needed to treat to avoid one peptic ulcer in elderly NSAID users was three, both in acute and chronic users [13].

Pantoprazole has been compared with *H. pylori* eradication therapy for the prevention of NSAID-related gastroduodenal damage. In a randomized study (n = 127), the treatment with pantoprazole (40 mg/day) for 1 month was found to be more effective than a PPI-based triple therapy in the prevention of gastroduodenal damage in elderly *H. pylori*-positive NSAID users [14].

Long-Term Treatment with Proton Pump Inhibitors (>3 Months of Nonsteroidal Anti-Inflammatory Drug Exposure)

Proton Pump Inhibitors versus Placebo

Controlled studies comparing PPIs and placebo for the prevention of tNSAID-induced gastroduodenal lesions varies from 3 to 12 months. A parallel group study (n = 169) compared omeprazole (20 mg/day) with placebo, with ongoing tNSAID treatment for 6 months [15]. In the placebo group, 16.5% of patients developed ulcers compared to 3.6% in patients receiving omeprazole. The estimated probability of remaining free of gastroduodenal lesions for patients taking omeprazole was 0.78 compared to 0.53 for the placebo ($p = 0.004$). In a larger randomized study (n = 175), omeprazole (20 mg/day) was compared with placebo in patients who needed continuous tNSAID treatment [16]. Although baseline endoscopy was not recorded, endoscopic evaluation at 3 months revealed that 4.7% of omeprazole-treated patients developed peptic ulcers, compared with 16.7% of the patients treated with placebo ($p < 0.001$).

In a randomized, parallel group study (n = 95), pantoprazole (40 mg/day) was compared with placebo for 12 weeks in patients with rheumatoid arthritis or osteoarthritis requiring tNSAIDs [17]. Baseline endoscopy confirmed gastric and duodenal lesions using the Lanza classification (grade 0: normal to hyperemic mucosa; grade 1: 1–3 erosions, submucosal hemorrhages or petechiae; grade 2: 4–10 erosions, submucosal hemorrhages or petechiae; grade 3: presence of >10 erosions, submucosal hemorrhages or petechia; grade 4: presence of an ulcer). After follow-up endoscopic evaluations, 77 and 59% of patients (those with a Lanza score of 0, 1, or 2 at base line) remained in remission after 12 weeks in the pantoprazole and placebo groups, respectively. There was a greater significant difference between the two groups in endoscopic remission, when patients had normal mucosa (grade 0) at base line (82% pantoprazole vs. 55% placebo, $p = 0.036$).

The largest of the randomized studies were VENUS (Verification of Esomeprazole for NSAID Ulcers and Symptoms) and PLUTO (Prevention of Latent Ulceration Treatment Options), both of which had similar study designs but with variation in locations (n = 844 and 585, respectively) [18]. VENUS was conducted in the USA, while PLUTO was a multinational study in 11 countries. Over a 6-month period, PPI cotherapy using esomeprazole (20 or 40 mg/day) was compared with placebo in patients requiring NSAIDs, including COX-2 inhibitors. Study patients were restricted to those with high risk for developing ulcers such as older age >60 years, past history of ulcers, and normal endoscopy at baseline. This differed from previous studies where patients with healed ulcers were recruited into the trial [6, 11]. Data from both studies showed that in chronic NSAID users with high GI risk factors, more patients remained ulcer free with omeprazole than with placebo. In the VENUS study, the proportion of patients developing ulcers was 4.7, 5.3 and 20.4% while on esomeprazole (40 mg/day, $p < 0.0001$), esomeprazole (20 mg/day, $p < 0.001$), and placebo, respectively. In the PLUTO study, the results were 4.4% with esomeprazole (40 mg/day, $p = 0.007$), 5.2% with esomeprazole (20 mg/day, $p = 0.018$), and 12.3% with placebo.

Thus, there is significant evidence to support the concomitant treatment with a PPI in the prevention of tNSAID-induced lesions. Further support comes from a meta-analysis showing that PPIs significantly reduced the risk of both endoscopic duodenal [relative risk (RR) = 0.39; 95% CI: 0.10–0.39)] and gastric ulcers (RR = 0.39; 95% CI: 0.31–0.50) when compared with placebo [19].

Proton Pump Inhibitors versus H_2-Receptor Antagonists

A large randomized study (n = 541), ASTRONAUT (Acid Suppression Trial: Ranitidine versus Omeprazole for NSAID-Associated Ulcer Treatment), compared omeprazole (20 and 40 mg/day) and ranitidine (120 mg twice daily) in both the healing and prevention of tNSAID-induced ulcers [6]. Patients selected were on continuous treatment with tNSAIDs and had endoscopic confirmation of ulcers or more than 10 erosions in the stomach or duodenum. Patients who had a successful treatment were recruited into the maintenance phase. During a maintenance therapy of 6 months, 72% of the patients in the omeprazole group (20 mg per day) remained in remission compared to 59% in the ranitidine group (RR = 0.32; 95% CI: 0.17–0.62). Further sub-analysis of patients with ulcers showed that those with smaller ulcers at initial endoscopy were more likely to have successful treatment at the healing phase ($p = 0.003$) and to continue remission during the maintenance phase ($p = 0.008$) than patients with larger ulcers. With the analysis of the ulcer site, there was an increased protection against duodenal ulcers with omeprazole maintenance therapy. Subsequent studies have not shown similar results. In a smaller randomized controlled trial (n = 194), lansoprazole (15 mg/day) was compared with famotidine (20 mg twice daily) involving patients with rheumatoid arthritis taking long-term NSAIDs. The study indicated no significant differences between the two groups [20].

Proton Pump Inhibitors versus Misoprostol

Although many studies have assessed the long-term effect of misoprostol on the prevention of tNSAID-induced ulcers, there are very few head-to-head studies comparing misoprostol with PPIs. One of the largest studies was the OMNIUM study (Omeprazole versus Misoprostol for NSAID-Induced Ulcer Management; n = 935) [11]. This double-blind randomized study had both a treatment phase and maintenance phase where omeprazole (20 mg/day) was compared with misoprostol (400 µg/day) or placebo for 6 months. There was a higher proportion of patients remaining in remission with maintenance treatment of omeprazole (61%) compared with misoprostol (48%, $p = 0.001$) or placebo (27%, $p < 0.001$). In subanalysis of the ulcer site, omeprazole was shown to be superior to misoprostol at reducing the risk of duodenal, but not gastric, ulcers.

Another randomized parallel group study (n = 515) compared the efficacy and the tolerability of pantoprazole (20 mg per day) with misoprostol (400 µg daily) in preventing tNSAID-related GI lesions. Endoscopy was performed at baseline, 3 months, and 6 months. Pantoprazole was superior to misoprostol: the remission rates at 6 months were 95% for pantoprazole and 86% for misoprostol ($p < 0.001$). A higher proportion of patients in the misoprostol group discontinued the study due to adverse effects (13% misoprostol vs. 5% pantoprazole) [21].

When a higher dose of misoprostol was used, it was shown to be superior to lansoprazole for prevention of NSAID-induced gastric ulcers [22]. This study also included patients taking low-dose aspirin, but excluded patients in whom aspirin use was greater than 1,300 mg per day. Patients using long-term tNSAID/aspirin (n = 637) were randomized to receive high-dose misoprostol (800 µg/day), lansoprazole (15 or 30 mg per day), or placebo. After 12 weeks, the percentage of gastric ulcer-free patients was 51% with placebo, 93% with misoprostol, 80% with lansoprazole 15 mg, and 82% with lansoprazole 30 mg. However, there was a much higher proportion of patients with adverse effects in the misoprostol group; therefore, when adjusted to withdrawal rates, there were no differences between the active treatment groups. Thus, due to lack of randomized trials with head-to-head comparison of PPI and misoprostol, a meta-analysis failed to reach any statistical difference between the two strategies [19].

Proton Pump Inhibitors and Prevention of Aspirin-Induced Gastroduodenal Lesions

Proton Pump Inhibitors versus Placebo

There is a lack of trial data on the effects of PPIs in aspirin users who do not have a history of ulcer bleeding. More published data exists on the concomitant use of PPI and tNSAIDs compared to that of aspirin. In the earlier studies, the protective effect of lansoprazole (30 mg/day) against aspirin-induced gastric lesions was

Table 2. Randomized controlled trials with PPIs as preventive therapy of aspirin-induced lesions; head-to-head comparison with H_2-RA and placebo

Author and year	Intervention	Study size, n	Follow-up weeks	Patients with GI lesions, %	p value
1994 [26]	omeprazole (30 mg o.d.)	20	2	15	
	placebo			30	<0.01
2002 [27]	lansoprazole (30 mg o.d.)	123	74	1.6	
	placebo			14.8	0.006
2008 [6]	esomeprazole (20 mg o.d.)	991	26	1.6	
	placebo			5.4	0.007
2010 [28]	pantoprazole (20 mg o.d.)	130	48	0	
	famotidine (20 mg o.d.)			12.3	0.0031

o.d. = Once daily.

shown in a double-blind crossover trial measuring the functional capacity of the mucosal barrier (by recording of the gastric potential difference) and of the morphological changes in the mucosa by endoscopy [23]. Further studies using PPIs for prevention of gastroduodenal ulcers after low-dose aspirin use were carried out in Germany. These two small randomized studies (n = 36 and 30) compared PPIs with placebo in healthy volunteers taking low-dose aspirin for 14 days. Both studies suggested that omeprazole and lansoprazole significantly reduced the risk of gastroduodenal lesions induced by aspirin (300 g daily) [24, 25]. In another short-term study (n = 20) in which a PPI was coprescribed with aspirin for 2 weeks, omeprazole (40 mg/day) was also shown to be effective in preventing aspirin-induced gastroduodenal damage [26]. Omeprazole protected 85% of the subjects from extensive gastric erosions (often associated with evidence of intraluminal bleeding) or ulceration, whereas 70% of the subjects developed aspirin-induced grades 3 and 4 gastric injury while on placebo ($p < 0.01$). No subject taking omeprazole developed duodenal injury of any grade, while 50% taking placebo developed erosions and 15% had ulcers ($p < 0.001$). Further larger studies supported their results. A randomized study (n = 123) recruited patients who had experienced an ulcer complication after low-dose aspirin. After the ulcer healed, the patients were randomized to lansoprazole (30 mg/day) or placebo while remaining on aspirin. During a median follow-up of 12 months, 14.8% in the placebo group had ulcer recurrence compared to 1.6% in the lansoprazole group (adjusted hazard ratio = 9.6; 96% CI: 1.2–76.1) [27].

The largest of the randomized controlled trials was by Yeomans et al. [6] (n = 991), in which esomeprazole (20 mg per day) was compared with placebo in patients receiving aspirin (75–325 mg once daily). Endoscopic assessment was carried out at baseline, week 8, and week 26. During the 26 weeks of treatment, 1.6% of the patients developed gastric or duodenal ulcers compared to 5.4% in the placebo group. The corresponding life-table estimates were 1.8 and 6.2% over 6 months, respectively ($p = 0.0007$), which equates to a relative reduction of developing an ulcer of 71% when taking esomeprazole rather than placebo.

Proton Pump Inhibitors versus H_2-Receptor Antagonists
Other possible options for secondary prevention of low-dose aspirin-induced gastroduodenal lesions include cotreatment with H_2-receptors, eradication of *H. pylori* infection, and substitution with clopidogrel. A randomized prospective study (n = 130) compared pantoprazole (20 mg/day) with famotidine (40 mg twice daily) in prevention of aspirin-related peptic ulcers. Recurrence of gastroduodenal lesions was more common in the famotidine group than with pantoprazole group (12.3 vs. 0%, $p = 0.0031$) [28].

Proton Pump Inhibitors and Prevention of Cyclooxygenase-2 Inhibitor-Induced Gastroduodenal Lesions

The GI adverse events associated with NSAIDs are thought to be due to inhibition of the COX-1 isoform; therefore, it was suggested that the selective inhibition of COX-2 might reduce their incidence while maintaining the therapeutic efficacy. This triggered a rise in the prescriptions of COX-2 inhibitors in the early 2000s, with a drop in sales of tNSAIDs. In 2004, however, further change in the usage of NSAID emerged after reports of cardiovascular toxicity caused by COX-2 inhibitors [1]. Many of the COX-2 inhibitors were removed from the market due to their cardiovascular side effects. However, COX-2 inhibitors remain on the market, and on the basis of GI safety alone, a systematic review demonstrated a clear advantage of COX-2 inhibitors over tNSAIDs [29]. In the same meta-analysis, it was shown that the addition of aspirin to COX-2 inhibitors significantly increased the risk of GI toxicity (perforation, obstruction, bleeding) – by more than fourfold over COX-2 inhibitors alone [29]. The addition of aspirin to a tNSAID demonstrated a nonsignificant 1.27-fold increased risk of GI toxicity over the use of a tNSAID alone.

Most data on concomitant treatment of PPI and COX-2 inhibitors comes from subgroup analysis of patients using COX-2 inhibitors. In a large retrospective cohort study (n = 1,383) comparing different gastroprotective strategies, a COX-2 inhibitor plus PPI strategy was associated with the highest magnitude of risk reduction for all upper GI complications (OR = 0.36; 95% CI: 0.28–0.47) [30]. Few randomized studies exist investigating the benefit of PPI and COX-2 inhibitors. In the few existing studies, they have focused on recurrent ulcer bleeding as study endpoints in high-risk patients rather than using follow-up endoscopy to check for gastroduodenal lesions.

The VENUS study reported pooled ulcer rates for patients using COX-2 inhibitors (n = 400) of 15.5% on placebo, 0.9% on esomeprazole 20 mg ($p < 0.001$), and 4.1% on esomeprazole 40 mg (p = 0.002) [18]. There is only one randomized study that investigates whether combined treatment of a COX-2 inhibitor (celecoxib) and PPI (esomeprazole) would be better than the COX-2 inhibitor alone in preventing recurrent ulcer bleeding [31]. After 13 months of follow-up (n = 441), the incidence of recurrent ulcer bleeding was significantly greater in patients who received celecoxib alone (8.9%) than those who received combination treatment (0%, p = 0.004).

Cotreatment of PPI and tNSAID has been compared with COX-2 inhibitors alone to see which analgesic strategy will give the least GI toxicity. A meta-analysis of six randomized controlled trials with a total of 6,219 patients revealed that there was no difference in the upper GI adverse events between COX-2 inhibitors and tNSAIDs with concurrent use of PPIs (RR = 0.61; 95% CI: 0.34–1.09) [32]. No studies exist to assess whether a COX-2 inhibitor with a PPI provide greater GI protection than a nonselective tNSAID with a PPI.

Review and Recommendation Based on Risk Assessment

The selection of analgesia as a tNSAID or COX-2 inhibitor has to be considered as well as prophylaxis use of PPIs for preventing NSAID-induced ulcers. The clinician's choice should be preceded by a prior assessment of the patient's cardiovascular and GI risk factors. Further confusion in the prescription of NSAIDs comes from reports of COX-2 inhibitors and higher doses of nonselective/non-naproxen NSAIDs sharing similar cardiovascular risks [1]. Therefore, in patients with average cardiovascular risk, there may be little advantage of using a nonselective/non-naproxen NSAID with a gastroprotective agent such as a PPI over the use of COX-2 inhibitors alone. For patients with high risk for GI complications, but with low cardiovascular risk factors, a PPI and a low-dose COX-2 inhibitor may be appropriate. In those with high cardiovascular risk factors, a nonselective NSAID combined with a PPI may be more beneficial.

References

1 Kearney PM, Baigent C, Godwin J, Halls H, Emberson JR, Patrono C: Do selective cyclo-oxygenase-2 inhibitors and traditional non-steroidal anti-inflammatory drugs increase the risk of atherothrombosis? Meta-analysis of randomised trials. BMJ 2006;332: 1302–1308.

2 Musumba C, Pritchard DM, Pirmohamed M: Review article: cellular and molecular mechanisms of NSAID-induced peptic ulcers. Aliment Pharmacol Ther 2009;30:517–531.

3 Flower R, Gryglewski R, Herbaczynska-Cedro K, Vane JR: Effects of anti-inflammatory drugs on prostaglandin biosynthesis. Nat New Biol 1972;238:104–106.

4 Sostres C, Lanas A: Gastrointestinal effects of aspirin. Nat Rev Gastroenterol Hepatol 2011;8:385–394.

5 Johansson C, Aly A, Befrits R, Smedfors B, Uribe A: Protection of the gastroduodenal mucosa by prostaglandins. Scand J Gastroenterol Suppl 1985;110:41–48.

6 Yeomans ND, Tulassay Z, Juhasz L, Racz I, Howard JM, van Rensburg CJ, et al: A comparison of omeprazole with ranitidine for ulcers associated with nonsteroidal antiinflammatory drugs. Acid Suppression Trial: Ranitidine versus Omeprazole for NSAID-Associated Ulcer Treatment (ASTRONAUT) study group. N Engl J Med 1998;338:719–726.

7 Agrawal NM, Campbell DR, Safdi MA, Lukasik NL, Huang B, Haber MM: Superiority of lansoprazole vs ranitidine in healing nonsteroidal anti-inflammatory drug-associated gastric ulcers: results of a double-blind, randomized, multicenter study. NSAID-associated gastric ulcer study group. Arch Intern Med 2000;160:1455–1461.

8 Goldstein JL, Johanson JF, Suchower LJ, Brown KA: Healing of gastric ulcers with esomeprazole versus ranitidine in patients who continued to receive NSAID therapy: a randomized trial. Am J Gastroenterol 2005; 100:2650–2657.

9 Goldstein JL, Johanson JF, Hawkey CJ, Suchower LJ, Brown KA: Clinical trial: healing of NSAID-associated gastric ulcers in patients continuing NSAID therapy – a randomized study comparing ranitidine with esomeprazole. Aliment Pharmacol Ther 2007;26:1101–1111.

10 Bianchi Porro G, Lazzaroni M, Petrillo M, Manzionna G, Montrone F, Caruso I: Prevention of gastroduodenal damage with omeprazole in patients receiving continuous NSAIDs treatment. A double blind placebo controlled study. Ital J Gastroenterol Hepatol 1998;30:43–47.

11 Hawkey CJ, Karrasch JA, Szczepanski L, Walker DG, Barkun A, Swannell AJ, et al: Omeprazole compared with misoprostol for ulcers associated with nonsteroidal antiinflammatory drugs. Omeprazole versus misoprostol for NSAID-induced ulcer management (OMNIUM) study group. N Engl J Med 1998;338: 727–734.

12 Labenz J, Blum AL, Bolten WW, Dragosics B, Rosch W, Stolte M, et al: Primary prevention of diclofenac associated ulcers and dyspepsia by omeprazole or triple therapy in *Helicobacter pylori* positive patients: a randomised, double blind, placebo controlled, clinical trial. Gut 2002;51:329–335.

13 Pilotto A, Franceschi M, Leandro G, Paris F, Cascavilla L, Longo MG, et al: Proton-pump inhibitors reduce the risk of uncomplicated peptic ulcer in elderly either acute or chronic users of aspirin/non-steroidal anti-inflammatory drugs. Aliment Pharmacol Ther 2004;20:1091–1097.

14 Pilotto A, Di Mario F, Franceschi M, Leandro G, Battaglia G, Germana B, et al: Pantoprazole versus one-week *Helicobacter pylori* eradication therapy for the prevention of acute NSAID-related gastroduodenal damage in elderly subjects. Aliment Pharmacol Ther 2000;14:1077–1082.

15 Cullen D, Bardhan KD, Eisner M, Kogut DG, Peacock RA, Thomson JM, et al: Primary gastroduodenal prophylaxis with omeprazole for non-steroidal anti-inflammatory drug users. Aliment Pharmacol Ther 1998;12:135–140.

16 Ekstrom P, Carling L, Wetterhus S, Wingren PE, Anker-Hansen O, Lundegardh G, et al: Prevention of peptic ulcer and dyspeptic symptoms with omeprazole in patients receiving continuous non-steroidal anti-inflammatory drug therapy. A Nordic multicentre study. Scand J Gastroenterol 1996;31:753–758.

17 Bianchi Porro G, Lazzaroni M, Imbesi V, Montrone F, Santagada T: Efficacy of pantoprazole in the prevention of peptic ulcers, induced by non-steroidal anti-inflammatory drugs: a prospective, placebo-controlled, double-blind, parallel-group study. Dig Liver Dis 2000;32:201–208.

18 Scheiman JM, Yeomans ND, Talley NJ, Vakil N, Chan FK, Tulassay Z, et al: Prevention of ulcers by esomeprazole in at-risk patients using non-selective NSAIDs and COX-2 inhibitors. Am J Gastroenterol 2006;101:701–710.

19 Rostom A, Dube C, Wells G, Tugwell P, Welch V, Jolicoeur E, et al: Prevention of NSAID-induced gastroduodenal ulcers. Cochrane Database Syst Rev 2002;4:CD002296.

20 Miyake K, Ueki N, Suzuki K, Shinji Y, Kusunoki M, Hiratsuka T, et al: Preventive therapy for non-steroidal anti-inflammatory drug-induced ulcers in Japanese patients with rheumatoid arthritis: the current situation and a prospective controlled-study of the preventive effects of lansoprazole or famotidine. Aliment Pharmacol Ther 2005;21(suppl 2):67–72.

21 Stupnicki T, Dietrich K, Gonzalez-Carro P, Straszak A, Terjung A, Thomas KB, et al: Efficacy and tolerability of pantoprazole compared with misoprostol for the prevention of NSAID-related gastrointestinal lesions and symptoms in rheumatic patients. Digestion 2003;68:198–120.

22 Graham DY, Agrawal NM, Campbell DR, Haber MM, Collis C, Lukasik NL, et al: Ulcer prevention in long-term users of nonsteroidal anti-inflammatory drugs: results of a double-blind, randomized, multicenter, active- and placebo-controlled study of misoprostol vs lansoprazole. Arch Intern Med 2002;162: 169–175.

23 Bergmann JF, Chassany O, Simoneau G, Lemaire M, Segrestaa JM, Caulin C: Protection against aspirin-induced gastric lesions by lansoprazole: simultaneous evaluation of functional and morphologic responses. Clin Pharmacol Ther 1992;52:413–416.

24 Simon B, Elsner H, Muller P: Protective effect of omeprazole against low-dose acetylsalicylic acid. Endoscopic controlled double-blind study in healthy subjects (in German). Arzneimittelforschung 1995; 45:701–703.

25 Müller P, Fuchs W, Simon B: Studies on the protective effect of lansoprazole on human gastric mucosa against low-dose acetylsalicylic acid. An endoscopic controlled double-blind study (in German). Arzneimittelforschung 1997;47:758–760.

26 Scheiman JM, Behler EM, Loeffler KM, Elta GH: Omeprazole ameliorates aspirin-induced gastroduodenal injury. Dig Dis Sci 1994;39:97–103.
27 Lai KC, Lam SK, Chu KM, Wong BC, Hui WM, Hu WH, et al: Lansoprazole for the prevention of recurrences of ulcer complications from long-term low-dose aspirin use. N Engl J Med 2002;346:2033–2038.
28 Ng FH, Wong SY, Lam KF, Chu WM, Chan P, Ling YH, et al: Famotidine is inferior to pantoprazole in preventing recurrence of aspirin-related peptic ulcers or erosions. Gastroenterology 2010;138:82–88.
29 Rostom A, Muir K, Dube C, Jolicoeur E, Boucher M, Joyce J, et al: Gastrointestinal safety of cyclooxygenase-2 inhibitors: a Cochrane Collaboration Systematic Review. Clin Gastroenterol Hepatol 2007;5:818–828, 828.e1–e5, quiz 768.
30 Targownik LE, Metge CJ, Leung S, Chateau DG: The relative efficacies of gastroprotective strategies in chronic users of nonsteroidal anti-inflammatory drugs. Gastroenterology 2008;134:937–944.
31 Chan FK, Wong VW, Suen BY, Wu JC, Ching JY, Hung LC, et al: Combination of a cyclo-oxygenase-2 inhibitor and a proton-pump inhibitor for prevention of recurrent ulcer bleeding in patients at very high risk: a double-blind, randomised trial. Lancet 2007;369:1621–1626.
32 Wang X, Tian HJ, Yang HK, Wanyan P, Peng YJ: Meta-analysis: cyclooxygenase-2 inhibitors are no better than nonselective nonsteroidal anti-inflammatory drugs with proton pump inhibitors in regard to gastrointestinal adverse events in osteoarthritis and rheumatoid arthritis. Eur J Gastroenterol Hepatol 2011;23:876–880.

Prof. Francis K.L. Chan
Institute of Digestive Disease, The Chinese University of Hong Kong
Prince of Wales Hospital, New Territories
Hong Kong, SAR (China)
E-Mail fklchan@cuhk.edu.hk

Chiba T, Malfertheiner P, Satoh H (eds): Proton Pump Inhibitors: A Balanced View.
Front Gastrointest Res. Basel, Karger, 2013, vol 32, pp 59–67 (DOI: 10.1159/000350631)

Proton Pump Inhibitors: Key Ingredients in *Helicobacter pylori* Eradication Treatment

Toshiro Sugiyama

Department of Gastroenterology, Graduate School of Medicine and Pharmaceutical Sciences University of Toyama, and Comprehensive Cancer Center, University Hospital of Toyama, Toyama City, Japan

Abstract

The combination of a proton pump inhibitor (PPI) with two of the following three antibiotics, amoxicillin, clarithromycin, and metronidazole, has been reported to achieve a high eradication rate. These regimens are accepted as standard regimens worldwide. The rationale of PPI-based triple therapy has been established by many clinical trials and a large body of experimental evidence. PPIs have a pivotal effect in the eradication of *Helicobacter pylori* infection, both indirect effects and direct effects. PPIs have a weak antibacterial effect against *H. pylori* in vitro, antiurease activity, and anti-ATPase activity. In fact, clinical trials have demonstrated that PPIs alone have a weak eradication effect. Indirect synergistic effects of PPIs could be raising the eradication rate in PPI-based triple regimens. The most potent effects are to stabilize and to raise antibacterial effects of the combined antibiotics, especially clarithromycin and amoxicillin in the hostile gastric environment. These are key antibiotics in an eradication treatment of *H. pylori* infection. In addition, PPIs have a potent activity to concentrate the antibiotics via suppressing gastric juice. Thus, the indirect effect might be an important role to raise the eradication rate by PPI-based triple therapy in clinical aspects. PPIs have different metabolic pathways, which depend on the CYP enzymes in the liver. Therefore, polymorphism of CYP enzymes might influence eradication rates in PPI-based regimens. The pharmacological characteristics of each PPI including the time exhibiting acid suppressive effect, metabolic pathway, and bioavailability should be considered to achieve a higher eradication rate in PPI-based triple treatment.

Helicobacter pylori is recognized as the main etiological factor in the pathogenesis of histologic gastritis, gastric ulceration, duodenal ulceration, gastric MALT lymphoma, and gastric cancer. Treatment to eradicate *H. pylori* infection has been proven to prevent completely the relapse of gastric or duodenal ulceration and has changed the natural history of ulcer diseases [1–3]. Recent strong evidence has demonstrated that the eradication of *H. pylori* infection might prevent the development of gastric cancer [4]. In addition, the eradication has been proven to improve gastric MALT lymphoma

[5] and the histological remission has been preserved for a long time and resulted in achieving long overall survival up to 95% in 10-year follow-up [6].

Numerous pharmacological regimens have been proposed for the eradication of *H. pylori* infection. The combination of a proton pump inhibitor (PPI) with two of the following three antibiotics, amoxicillin, clarithromycin and nitroimidazole, has been reported to achieve a high eradication rate. These regimens led to the acceptance of PPI-based triple therapy in the recommendation of the guidelines in Europe [7] and Japan [8]. The rationale of PPI-based triple therapy has been established by many clinical trials [9] as well as in vitro experimental data. In this chapter, the role of PPIs in eradication treatment will be discussed in terms of clinical and basic aspects.

Proton Pump Inhibitors Are Key Ingredients in Eradication Regimens

Early trials combining acid suppression drugs with antibiotics were a quadruple therapy with ranitidine and triple therapy with bismuth, metronidazole, and tetracycline [10]. Inhibition of acid secretion by a PPI, omeprazole, given twice a day in combination with amoxicillin has achieved eradication rates as high as those seen with classical triple therapy [11]. Classical triple therapy was associated with problems of compliance and significant side effects in more than 30% of patients, which included dizziness, nausea, diarrhea, colitis, metallic taste, and resistance to metronidazole. To confirm the additive role of PPIs in eradication treatment, four large-scale multicenter clinical trials, the MACH1 [12], MACH2 [13], GU-MACH [14], and DU-MACH studies [15], were conducted.

In the MACH2 study [13], the eradication rates of four randomized groups, the OAC group (omeprazole: 20 mg, amoxicillin: 1,000 mg and clarithromycin: 500 mg, twice daily) versus the AC group (amoxicillin: 1,000 mg and clarithromycin: 500 mg, twice daily) and the OMC group (omeprazole: 20 mg, metronidazole: 400 mg, and clarithromycin: 250 mg, twice daily) versus the MC group (metronidazole: 400 mg and clarithromycin: 250 mg, twice daily), were compared. The eradication rates for the OAC and AC groups for 7 days were 94% (n = 127) and 26% (n = 131), respectively, in intention-to-treat (ITT) analysis, and 95% (n = 110) and 25% (n = 110), respectively, in per-protocol analysis. On the other hand, the eradication rates of the OMC and MC groups for 7 days were 87% (n = 127) and 69% (n = 129), respectively, in ITT analysis, and 91% (n = 116) and 72% (n = 113), respectively, in per-protocol analysis (fig. 1). The eradication rates against clarithromycin-resistant strains were 100% (n = 2) in the OAC group and 0% (n = 1) in the AC group. The eradication rates against metronidazole-resistant strains were 76% (n = 25) in the OMC group and 43% (n = 15) in the MC group. The additive eradication effect of PPIs was higher with the regimen of the AC group than with the regimen of the MC group. However, the frequency of common adverse events, diarrhea, taste perversion, and liver dysfunction was equal for the OAC and AC groups as well as the OMC and MC groups. Therefore,

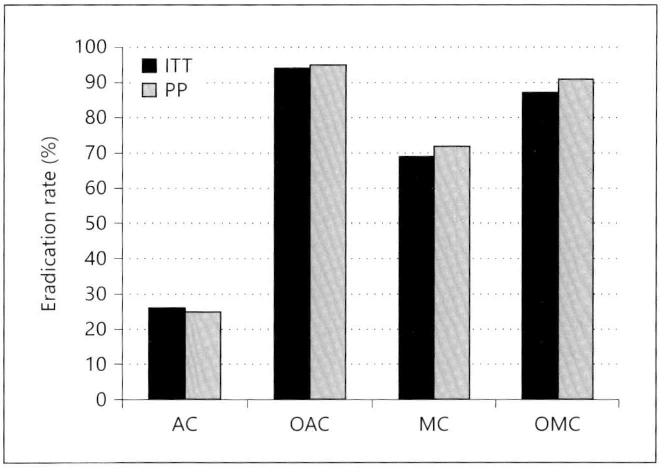

Fig. 1. PPIs are essential for eradication treatment (modified from [13]). AC = Amoxicillin: 1,000 mg and clarithromycin: 500 mg, twice daily; OAC = omeprazole: 20 mg, amoxicillin: 1,000 mg, and clarithromycin: 500 mg, twice daily; MC = metronidazole: 400 mg and clarithromycin: 250 mg, twice daily; OMC = omeprazole: 20 mg, metronidazole: 400 mg, and clarithromycin: 250 mg, twice daily; PP = per-protocol analysis.

the addition of omeprazole to the regimens of the AC or MC groups achieved a very high eradication rate and was able to overcome primary resistance to clarithromycin- or metronidazole-resistant strains.

In the GU-MACH2 study [14], the efficacy of omeprazole itself in the eradication of *H. pylori* infection was studied. The eradication rates of the OAC group (omeprazole: 20 mg, amoxicillin: 1,000 mg, and clarithromycin: 500 mg, twice daily) and the OMC group (omeprazole 20 mg, clarithromycin 250 mg, and metronidazole 400 mg, twice daily) for 7 days were 79 and 86%, respectively, in ITT analysis. However, the eradication rate of the O group (omeprazole: 20 mg, twice daily) was only 4%. This result suggested that PPIs themselves have a weak antibacterial effect against *H. pylori* and the potent effect in eradication treatment might be due to the synergistic effects to amoxicillin and clarithromycin, or metronidazole and clarithromycin.

How Do Proton Pump Inhibitors Show Antibacterial Effects against *H. pylori?*

PPIs have a pivotal effect in the eradication of *H. pylori* infection, both direct and indirect effects (table 1). PPIs have a weak antibacterial activity against *H. pylori* in vitro at an MIC_{50} between 12.5 and 50 μg/l [16, 17]. This effect can be enhanced by the substitution of the fluoroalkoxy group at position 4 of the pyrimidine ring of the PPIs. Additionally, several acid conversion metabolites of PPIs have also shown a more potent antibacterial effect than the parent PPIs. In addition, the in vitro inhibition of *H. pylori* urease by PPIs is speculated to result from the covalent binding by the substituted benzimidazoles to sulfhydryl groups in urease of *H. pylori* [18]. The third direct effect is inhibition of ATPase of *H. pylori* [19]. All PPIs can bind to ATPase of the p type on membrane of the live *H. pylori* and can inhibit the enzyme activity in an acid environment. These effects may contribute to the survival of *H. pylori* under hostile

Table 1. Anti-*H. pylori* activity of PPIs

Indirect effects	Direct effects
Elevation of antibacterial activity of antibiotics in the stomach	Antibacterial activity
Concentration of gastric juice	Antiurease activity
Stabilization of sIgA in the gastric lumen	Anti-ATPase activity

Table 2. MIC_{90} of antimicrobial agents for *H. pylori* at varying pH (modified from [20])

MIC_{90}, µg/ml	pH 5.7–6.0	pH 7.4
Ampicillin	2	0.25
Erythromycin	>4	0.25
Ciprofloxacin	4	0.5
Ofloxacin	2	1
Metronidazole	4	4
Doxycycline	2	2

acidic environment in the stomach. However, these in vitro antibacterial effects against *H. pylori* are doubtful to contribute to the raising of eradication rates in vivo. As mentioned previously, PPIs alone definitely showed a weak eradication effect in several clinical trials.

An indirect synergistic effect of PPIs could be the raising of eradication rates in PPI-based triple regimens. The most potent effects are to stabilize and to raise antibacterial activities of the combined antibiotics. Several papers have demonstrated that the MIC of antibiotics against *H. pylori* is dependent on an intragastric pH (table 2) [20, 21]. The MIC_{90} of ampicillin, the active drug of amoxicillin, is 2 µg/ml under pH 5.7–6.0 and is 0.25 µg/ml under pH 7.4. In a neutral gastric pH, antibacterial activity against *H. pylori* increases to eight times. In erythromycin, the prototype of clarithromycin, antibacterial activity against *H. pylori* increases up to approximately 20 times or more. These are key antibiotics in an eradication treatment of *H. pylori* infection. However, the MIC_{90} of metronidazole against *H. pylori* is not dependent on intragastric pH.

As shown in the MACH2 study, the additive eradication effect of PPI with two antibiotics is different in the AC and MC groups. The OAC group demonstrated a higher additive effect of PPIs than the OMC group (70 vs. 19% increase). In addition, we have recently published that the eradication rate of PPI-based triple therapy containing metronidazole is equal to a strong histamine receptor inhibitor-based triple therapy in second-line treatment against *H. pylori* infection [22]. The second indirect

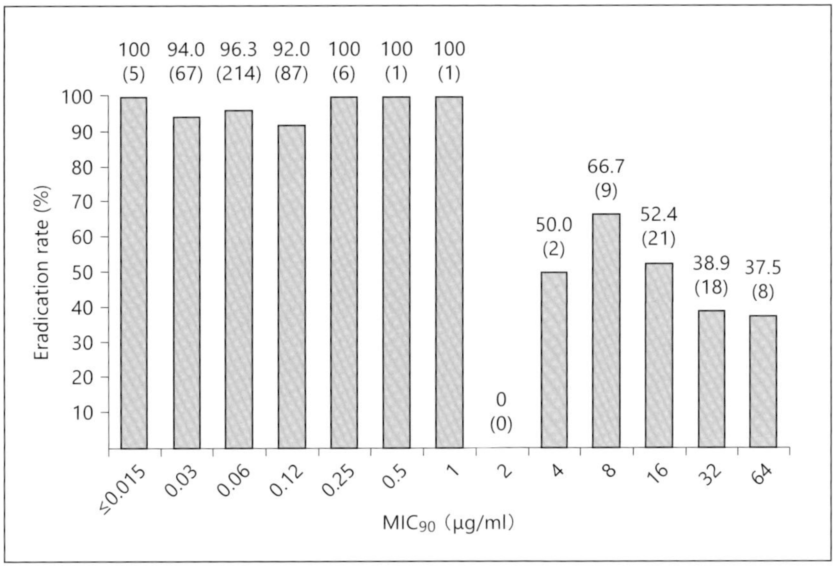

Fig. 2. Eradication rate of PPI-based triple therapy containing clarithromycin at the in vitro MIC level of clarithromycin (adapted from [24]). Numbers in parentheses represent sample number.

effect is the concentration of the combined antibiotics in the stomach. PPIs have a very strong inhibitory effect of gastric acid secretion, resulting in the concentration of antibiotics in the stomach. These two effects might contribute to elevate the anti-*H. pylori* activity in the hostile environment. The third effect is the stabilization of *H. pylori*-specific IgA secreted into gastric lumen [23]. IgA secreted into gastric lumen is rapidly degraded by the proteolytic activity of gastric juice. Although PPIs have a direct and an indirect antibacterial activity against *H. pylori*, the indirect effect might be important for raising the eradication rates by PPI-based triple therapy in clinical aspects.

What Proton Pump Inhibitors Are Desirable in Eradication Treatment?

As described previously, potent acid suppression activity is needed to achieve a higher eradication rate in PPI-based triple therapy against *H. pylori* infection because the antibacterial activity of the combined antibiotics has been raised by depending on intragastric pH. The anti-*H. pylori* activity of clarithromycin is strongly dependent on intragastric pH. In neutral pH, it increases to over 20-fold. This potent activity could reach to the MIC_{90} level of intermediate-resistant strains in the gastric tissues and gastric lumen, and could overcome clarithromycin resistance (partially) [24]. In fact, as shown in figure 2, intermediate-resistant strains against clarithromycin were eradicated by a PPI-based triple regimen containing clarithromycin. In a high-dose PPI triple regimen

Fig. 3. pH ≥3 holding time (%) at pretreatment, day 1, day 2, and day 3 after administration of PPIs. Dotted line indicates 30% of 24 h in pH ≥3 holding time (adapted from [24]). RAB = Rabeprazole; LAN = lansoprazole; OME = omeprazole.

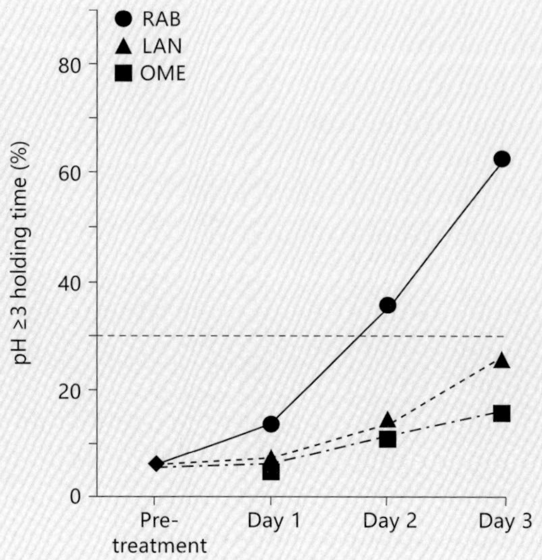

(double dose of PPIs, amoxicillin: 750 mg, and clarithromycin: 200 mg, twice daily), the eradication rate against clarithromycin-resistant strains was going-up approximately 10%. However, the eradication rate gradually decreased according to MIC level and failed in high-resistant strains. Five PPIs, omeprazole, lansoprazole, pantoprazole, rabeprazole, and esomeprazole, are available for eradication treatment. In general, the acid suppression activity of PPIs is approximately equal in clinical dosage of PPIs. However, as mentioned in other chapters, the metabolic pathway in the liver is different among PPIs due to polymorphism of the CYP enzyme, especially CYP2C19 [25]. In extensive metabolizers in liver, PPIs are cleared more rapidly from blood, resulting in lower plasma levels of the active form and less acid suppression activity. Thus, the polymorphism of CYP enzymes influence the eradication rates via various acid suppression activity of each PPI [26]. Omeprazole and lansoprazole have been shown to be metabolized by CYP2C19. Rabeprazole is metabolized in the liver by a nonenzymatic pathway, and esomeprazole (the optical isomer S-form of omeprazole) shows less clearance in the liver [27] and a weak dependency on the polymorphism of CYP enzymes.

The second influencing factor is the duration for exhibiting acid suppression after administration. The duration for eradication by PPI-based triple therapy is usually 7 days. The time to exhibit acid suppression after taking PPIs is critical. If it takes 3 days to reach a neutral pH, the actual effective duration for taking two antibiotics is 4 days. Saitoh et al. [28] reported that the duration to suppress an acid secretion after taking PPIs was different among PPIs (fig. 3). Rabeprazole and esomeprazole show acid suppression after 1 day and are faster than lansoprazole and omeprazole. In a clinical trial, the short duration has been proven to have an influence on the eradica-

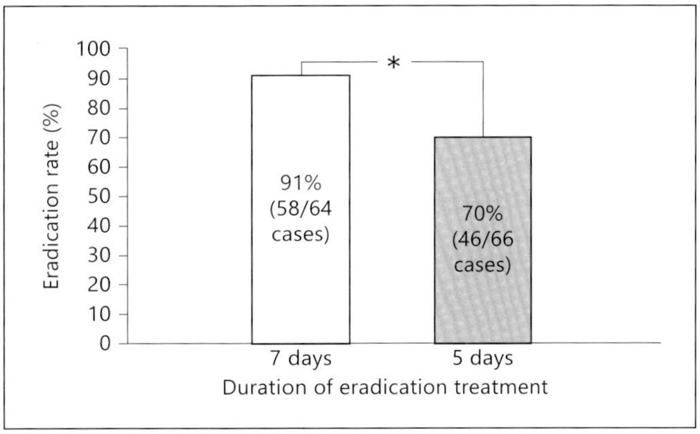

Fig. 4. Eradication rates with the same PPI-based triple therapy of 7-day or 5-day durations (modified from [29]). * $p < 0.01$.

tion rate. Isomoto et al. [29] compared the eradication rates of the same PPI-based triple regimens for 7-day administration and 5-day administration, with the latter showing a lower eradication rate (fig. 4). These results suggest that a stronger and faster acid suppressive activity of PPIs is required to achieve higher eradication rates against *H. pylori* infection for 7-day therapies. In a meta-analysis by Calvet et al. [30], 14-day PPI-based regimens are superior to 7-day PPI-based regimens in eradication treatment. Therefore, the pharmacological characteristics of each PPI, including the time of acid-suppressing effect, metabolic pathway, and bioavailability should be considered in regard to a higher eradication rate in PPI-based triple treatment.

What Regimen Is Recommended in the New Era?

At present, PPI-based triple regimens including clarithromycin are recommended as a first-line treatment. According to the increase of clarithromycin-resistant strains worldwide, the eradication rates are likely to fall below 80%. Second-line regimens vary among countries due to the unique antibiotic resistance in each country. Most countries, with the exception of Japan, demonstrate very high resistance rates against metronidazole. Therefore, ranitidine bismuth-based triple therapy (PPI, TC, MNZ) or levofloxacin-based quadruple therapy is recommended. A recent trend is sequential treatment, which was first reported in Italy. It consists of a PPI plus amoxicillin for 5 days first and followed by a PPI plus clarithromycin plus tinidazole for 5 days. The high eradication rate of this regimen has been demonstrated even in high-metronidazole-resistant countries, such as Korea, Italy, and other European countries. The second-line regimen recommended in the Japanese guideline is a PPI-based triple regimen containing metronidazole substituted for clarithromycin because of high clarithromycin resistance rates (39% in 2011). Nevertheless, first-line and second-line resistant strains are gradually increasing. A multicenter trial of a third-line regimen has recently been reported [31]. Two hundred and four patients who had unsuccessful treatments with first- and second-line eradication regimens were assessed in three randomized groups. The eradication rate of PPI-based triple therapy (lansoprazole: 30 mg, amoxicillin: 750 mg, and sitafloxacin: 100 mg, twice daily) for 7 days was 70.0% (n = 68) and that of high-dose PPI-based dual therapy (lansoprazole: 30 mg and amoxicillin: 500 mg, 4 times daily) for 14 days was 54% in crossover ITT analysis. Even if amoxicillin-resistant strains are very rare worldwide in first- and second-treatment failures, the eradication rate of high-dose PPI-based dual therapy is limited. Sitafloxacin is a novel quinolone that has an antibacterial activity in vitro against the resistant strains, which is induced by a *gyrA* mutation. Although it has been suggested that PPI-based triple therapy including STFX may be a candidate for a third-line eradication regimen, the eradication rate is insufficient. More studies including PPI selection, dosage, duration, and predicting factors for successful eradication may be required.

References

1 Hentschel E, Brandstatter G, Dragosics B, et al: Effect of ranitidine and amoxicillin plus metronidazole on the eradication of *Helicobacter pylori* and the recurrence of duodenal ulcer. Gastroenterology 1997;113:1082–1086.
2 Leodolter A, Kulig M, Brasch H, et al: A meta-analysis comparing eradication, healing and relapse rates in patients with *Helicobacter pylori*-associated gastric or duodenal ulcer. Aliment Pharmacol Ther 2001;15:1949–1958.
3 Miwa H, Sakaki N, Sugiyama T, et al: Recurrent peptic ulcers in patients following successful *Helicobacter pylori* eradication: a multicenter study of 4,940 patients. Helicobacter 2004;9:9–16.
4 Fukase K, Kato M, Kikuchi S, et al: Effect of eradication of *Helicobacter pylori* on incidence of metachronous gastric carcinoma after endoscopic resection of early gastric cancer: an open-label, randomised controlled trial. Lancet 2008;372:392–397.
5 Wotherspoon AC, Doglioni C, Diss TC, et al: Regression of primary low-grade B-cell gastric lymphoma of mucosa-associated lymphoid tissue type after eradication of *Helicobacter pylori*. Lancet 1993;342:575–577.
6 Nakamura S, Sugiyama T, Matsumoto T, et al: Long-term clinical outcome of gastric MALT lymphoma after eradication of *Helicobacter pylori*: a multi-center cohort follow-up study of 420 patients in Japan. Gut 2012;61:507–513.
7 Malfertheiner P, Megraud F, O'Morain C, et al: Current concepts in the management of *Helicobacter pylori* infection: the Maastricht III Consensus Report. Gut 2007;56:772–781.
8 Asaka M, Kato M, Sugiyama T, et al: Guidelines for the management of *Helicobacter pylori* infection in Japan: 2009 revised edition. Helicobacter 2010;15:1–20.
9 Ford AC, Delaney BC, Forman D, et al: Eradication therapy in *Helicobacter pylori* positive peptic ulcer disease: systematic review and economic analysis. Am J Gastroenterol 2004;99:1833–1855.
10 Logan RPH, Gummett PA, Misiewicz JJ, et al: One week's anti-*Helicobacter pylori* treatment for duodenal ulcer. Gut 1994;35:15–18.
11 Labenz J, Beker JA, Dekker CP, et al: Doubling the omeprazole dose (40 mg b.d. vs. 20 mg b.d.) in dual therapy with amoxicillin increases the cure rate of *Helicobacter pylori* infection in duodenal ulcer patients. Aliment Pharmacol Ther 1997;11:515–522.
12 Lind T, Veldhuyzen ZS, Unge P, et al: Eradication of *Helicobacter pylori* using one-week triple therapies combining omeprazole with two antimicrobials: the MACH1 study. Helicobacter 1996;1:138–144.
13 Lind T, Megraud F, Unge P, et al: The MACH2 study: role of omeprazole in eradication of *Helicobacter pylori* with 1-week triple therapies. Gastroenterology 1999;116:248–253.
14 Malfertheiner P, Bayerdörffer E, Diete U, et al: The GU-MACH study: the effect of 1-week omeprazole triple therapy on *Helicobacter pylori* infection in patients with gastric ulcer. Aliment Pharmacol Ther 1999;13:703–712.
15 Zanten SJ, Bradette M, Farley A, et al: The DU-MACH study: eradication of *Helicobacter pylori* and ulcer healing in patients with acute duodenal ulcer using omeprazole based triple therapy. Aliment Pharmacol Ther 1999;13:289–295.

16 Iwahi T, Satoh H, Nakao M: Lansoprazole, a novel benzimidazole proton pump inhibitor, and its related compounds have selective activity against *Helicobacter pylori*. Antimicrob Agents Chemother 1991; 35:490–496.
17 Mainguet P, Delmee M, Debongnie JP: Omeprazole, *Campylobacter pylori*, and duodenal ulcer. Lancet 1989;2:289–290.
18 Nagata K, Satoh H, Iwahi T, et al: Proton inhibitory action of the gastric proton pump inhibitor lansoprazole against urease activity of *Helicobacter pylori*: unique action selective for *H. pylori* cells. Antimicrob Agents Chemother 1993;37:769–774.
19 Mauch F, Bode G, Malfertheiner P: Identification and characterization of an ATPase system of *Helicobacter pylori* and effect of proton pump inhibitors. Am J Gastroenterol 1993;88:1801–1802.
20 Grayson ML, Elipoulos GM, Ferraro MJ, et al: Effect of varying pH on the susceptibility of *Campylobacter pylori* to antimicrobial agents. Eur J Clin Microbiol Infect Dis 1989;8:888–889.
21 McNulty CA, Dent JC, Ford GA, et al: Inhibitory antimicrobial concentrations against *Campylobacter pylori* in gastric mucosa. J Antimicrob Chemother 1988;22:729–738.
22 Kudo T, Fujinami H, Sugiyama T, et al: Comparison of lafutidine and rabeprazole in 7-day second-line amoxicillin- and metronidazole-containing triple therapy for *Helicobacter pylori*: a pilot study. Helicobacter 2012;17:277–281.
23 Sugiyama T, Furuyama S, Awakawa T, et al: Local immune response in gastric mucosa to *Helicobacter pylori* infection with and without intestinal metaplasia. Eur J Gastroenterol Hepatol 1993;5:119–122.
24 Kuwayama H, Asaka M, Sugiyama T, et al: Rabeprazole-based eradication therapy for *Helicobacter pylori*: a large-scale study in Japan. Aliment Pharmacol Ther 2007;25:1105–1113.
25 Furuta T, Shirai N, Ohashi K, et al: Therapeutic impact of CYP2C19 pharmacogenetics on proton pump inhibitor-based eradication therapy for *Helicobacter pylori*. Methods Find Exp Clin Pharmacol 2003;25: 131–143.
26 Hokari K, Sugiyama T, Kato M, et al: Efficacy of triple therapy with rabeprazole for *Helicobacter pylori* infection and CYP2C19 genetic polymorphism. Aliment Pharmacol Ther 2001;15:1479–1484.
27 Abelo A, Anderson TB, Antonsson M, et al: Stereoselective metabolism of omeprazole by human cytochrome P450 enzymes. Drug Metab Dispos 2000;28: 966–972.
28 Saitoh T, Fukushima Y, Otsuka H, et al: Effects of rabeprazole, lansoprazole and omeprazole on intragastric pH in CYP2C19 extensive metabolizers. Aliment Pharmacol Ther 2002;16:1811–1817.
29 Isomoto H, Furusu H, Morikawa T, et al: 5-day vs. 7-day triple therapy with rabeprazole, clarithromycin and amoxicillin for *Helicobacter pylori* eradication. Aliment Pharamacol Ther 2000;14:1619–1623.
30 Calvet X, Garcia N, Lopez T, et al: A meta-analysis of short versus long therapy with a proton pump inhibitor, clarithromycin and either metronidazole or amoxycillin for treating Helicobacter pylori infection. Aliment Pharmacol Ther 2000;14:603–609.
31 Murakami K, Furuta T, Ando T, et al: Multi-center randomized controlled study to establish the standard third-line regimen for *Helicobacter pylori* eradication in Japan. J Gastroenterol 2013, E-pub ahead of print.

Prof. Toshiro Sugiyama, MD, PhD
Department of Gastroenterology
Graduate School of Medicine and Pharmaceutical Sciences
University of Toyama, 2630 Sugitani, Toyama City (Japan)
E-Mail tsugi@med.u-toyama.ac.jp

Chiba T, Malfertheiner P, Satoh H (eds): Proton Pump Inhibitors: A Balanced View.
Front Gastrointest Res. Basel, Karger, 2013, vol 32, pp 68–76 (DOI: 10.1159/000350632)

Proton Pump Inhibitor Management in Bleeding Peptic Ulcer Disease

Joseph Sung

The Chinese University of Hong Kong, Hong Kong, SAR, China

Abstract

Proton pump inhibitors (PPIs) are effective in preventing recurrent peptic ulcer bleeding when used in high-dose intravenous infusion or oral form. Intermittent injection or low-dose administration of PPIs may not produce such benefit. These findings are based on meta-analysis that pooled multiple randomized control trials. On the other hand, PPIs given endoscopically may be able to reduce the need for endoscopic therapy in peptic ulcer disease. However, they have no proven value in reducing mortality.

Why Proton Pump Inhibitors in Bleeding Peptic Disease?

In a high acid milieu, platelet aggregation is severely impaired [1]. That is why bleeding from peptic ulcer disease and esophageal erosions from gastroesophageal reflux can be difficult to control. Indeed, gastric acid impairs clot formation and even favors platelet disaggregation [2, 3]. In order to regain platelet function, gastric acid secretion must be reduced and intragastric pH brought up to near neutrality (pH >6). It is impossible for an H_2-antagonist to achieve such potent acid suppressing function. Proton pump inhibitors (PPIs) are the ideal agents to restore the aggregating power of platelets.

In peptic ulcer bleeding, it is expected that PPIs may be able to consolidate the blood clot covering the ulcer and stabilize the condition. Because of this mechanism of action, one would expect the benefit of PPIs can only be seen when the ulcer is either actively bleeding or showing an adherent clot (or protuberant vessel with a sentinel clot). Low-risk ulcers with a clean base or just a pigmented spot do not need the pharmacologic action of PPIs for stabilization.

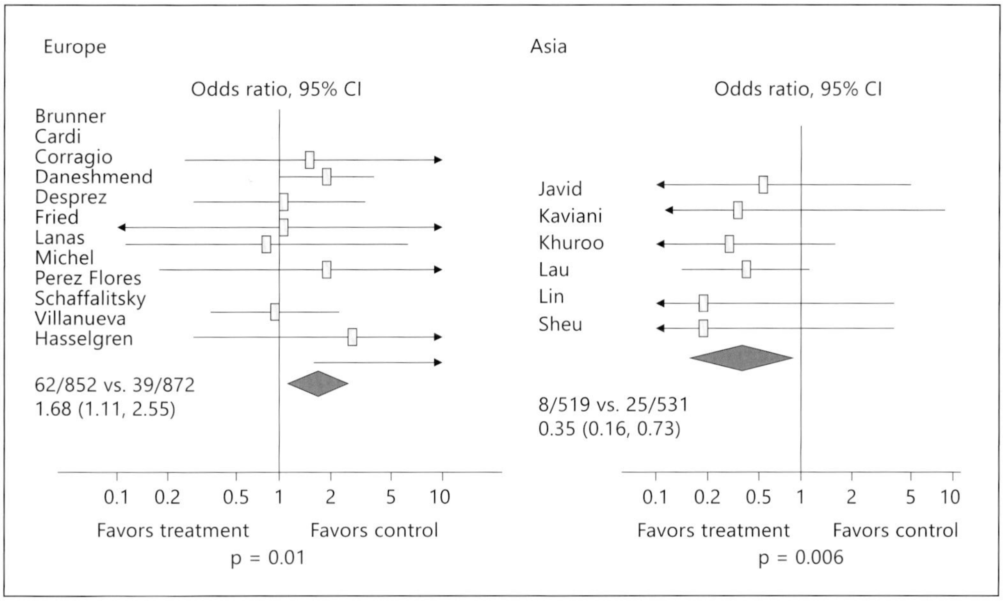

Fig. 1. Discrepant results of PPIs in peptic ulcer bleeding: recurrent bleeding [for details of the studies see 15].

Do Proton Pump Inhibitors Work in Real-Life Peptic Ulcer Bleeding?

Early studies have yielded conflicting results concerning the efficacy of intravenous PPIs in the management of acute upper gastrointestinal bleeding. While studies from Europe and North America have failed to demonstrate a benefit from PPIs [4–6], Asian studies have clearly demonstrated that PPIs reduce the risk of recurrent bleeding from peptic ulcer disease after endoscopic hemostasis [7, 8] (fig. 1). The discrepant results can be due to many factors, differences in patient choice, variations in outcome parameters, *Helicobacter pylori* infection, dosage and the administration of PPIs, etc., just to name a few.

In order to settle the dispute that intravenous PPIs do help prevent recurrent peptic ulcer bleeding, a multicenter multiethnic study was set up to address the issue [9]. Ninety-one hospital emergency departments in 16 countries joined together to recruit patients 18 years or older with peptic ulcer bleeding from a single gastric or duodenal ulcer showing high-risk stigmata. An intravenous esomeprazole bolus (80 mg) followed by an 8-mg/h infusion over 72 h or matching placebo were given after successful endoscopic hemostasis. Intervention was allocated by computer-generated randomization. After infusion, both groups received oral esomeprazole (40 mg/day) for 27 days. The primary endpoints included recurrent bleeding within 72 h, within 7 and 30 days, all-cause mortality, surgery, endoscopic retreatment, blood transfusions, and hospitalization. A total of 767 patients were randomly

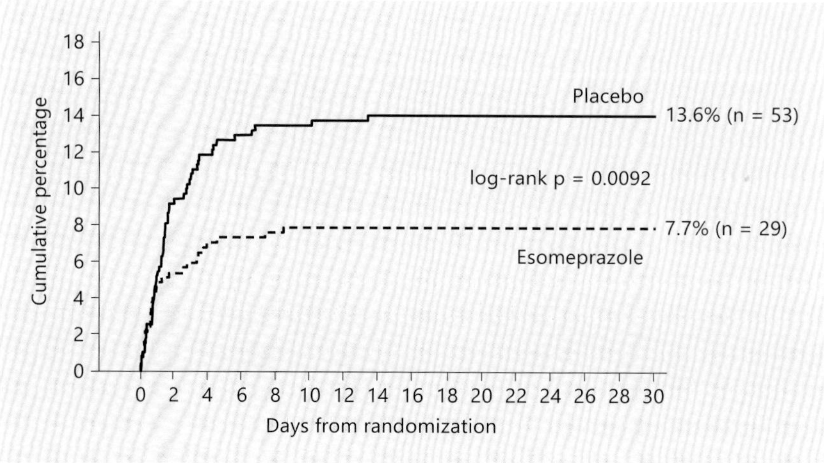

Fig. 2. Clinically significant recurrent bleeding within 30 days after endoscopic therapy in peptic ulcer bleeding [9].

assigned in this study and 764 provided data for an intention-to-treat analysis (375 esomeprazole recipients and 389 placebo recipients). Fewer patients receiving intravenous esomeprazole had recurrent bleeding within 72 h than those receiving placebo [5.9 vs. 10.3%; difference: 4.4 percentage points (95% CI: 0.6–8.3)]. The difference in bleeding recurrence remained significant at 7 and 30 days ($p < 0.010$). Esomeprazole also reduced endoscopic retreatment [6.4 vs. 11.6%; difference: 5.2 percentage points (95% CI: 1.1–9.2)], surgery (2.7 vs. 5.4%), and all-cause mortality rates (0.8 vs. 2.1%) more than placebo, although differences for the latter two comparisons were not significant.

The result of this study confirmed four important issues. First, there is no difference in the outcome between Europeans and Asians in their response to PPI therapy. Second, recurrent bleeding usually occurs within 72 h after the index bleeding; therefore, continuous therapy with intravenous PPIs for the first 3 days is justified (fig. 2). Third, the efficacy of PPIs in preventing recurrent bleeding is actually more remarkable in *Helicobacter pylori*-infected subjects (3.7 vs. 8.4%) compared to non-infected patients (9.8 vs. 11.4%). Finally, intravenous PPIs prevent recurrent bleeding after either single endoscopic therapy (epinephrine injection) or combined endoscopic therapy (epinephrine injection combined with hemoclips or thermal coagulation).

One may suspect that results from clinical trials under a controlled setting may not reflect the real-life situation. To assess clinical outcomes in patients treated with PPIs after endoscopic hemostasis in routine clinical care and to compare these outcomes to those seen in a randomized controlled trial (RCT) of intravenous esomeprazole, a study from Spain analyzed patients with peptic ulcer bleeding and endoscopic stigmata of recent hemorrhage who were treated with intravenous esomepra-

Table 1. Intermittent intravenous injection of PPIs does not prevent recurrent bleeding form peptic ulcer disease

Study	n	Treatment regimen		Rebleeding outcome
		omeprazole IV	control	
Daneshmend et al. [11]	1,147	80 + 40 mg, q8h × 3 doses	placebo	NS
Orti et al. [12]	519	80 + 40 mg, q8h for 48 h	RAN 50 mg IV, bolus + 50 mg q6h	NS
Villanueva et al. [13]	86	80 + 40 mg, q8h for 4 days	RAN 50 mg IV, q6h	NS
Lanas et al. [14]	51	80 + 40 mg, q12h	RAN 50 mg IV, q4h	NS

NS = Not significant; RAN = ranitidine.

zole or pantoprazole ≥120 mg/day following therapeutic endoscopy [10]. This study enrolled 539 patients from 12 hospitals in Spain. Overall, 9.1% of the patients experienced further bleeding within 72 h following initial endoscopy, 14.3% of the patients had further bleeding within 30 days, and 3.3% of the patients died within 30 days. In the RCT, the rate of recurrent bleeding within 72 h was significantly lower in the esomeprazole arm (5.9%) than in the placebo arm (10.3%). The recurrent bleeding rate in patients treated with esomeprazole in routine clinical practice (7.8%; 95% CI: 4.6–11.1) fell between these two values. Therefore, the benefit of intravenous esomeprazole was confirmed.

Would Intermittent Injection of Proton Pump Inhibitors or Low-Dose Proton Pump Inhibitors Work?

There are quite a number of studies, which were subsequently pooled into a meta-analysis, that show an intermittent dose of an intravenous PPI cannot produce the benefit as a bolus dose followed by continuous infusion (table 1) [11–14].

Furthermore, a low-dose infusion/injection of a PPI may also produce the same effect of preventing recurrent bleeding, as shown in a Cochrane review (fig. 3) [15]. However, as the Cochrane review pointed out, there is substantial heterogeneity in the many studies that were pooled together. A recent small-scale study from Taiwan showed that pantoprazole given as an infusion of 196 mg/day (10% recurrent bleeding) made no difference in the clinical outcome compared to intermittent injection of 40 mg every 6 h (8.4% recurrent bleeding) [16]. This data can be interpreted as either 196 and 160 mg per day are both high-dose administration, or that intravenous pantoprazole does not actually produce the same benefit as omeprazole or esomeprazole. A much larger study from China produced a quite different result. When 875 patients with peptic ulcer bleeding were randomized to receive either the standard regimen, consisting of an intravenous bolus of 40 mg of PPI twice a day with a continuous infusion of saline for

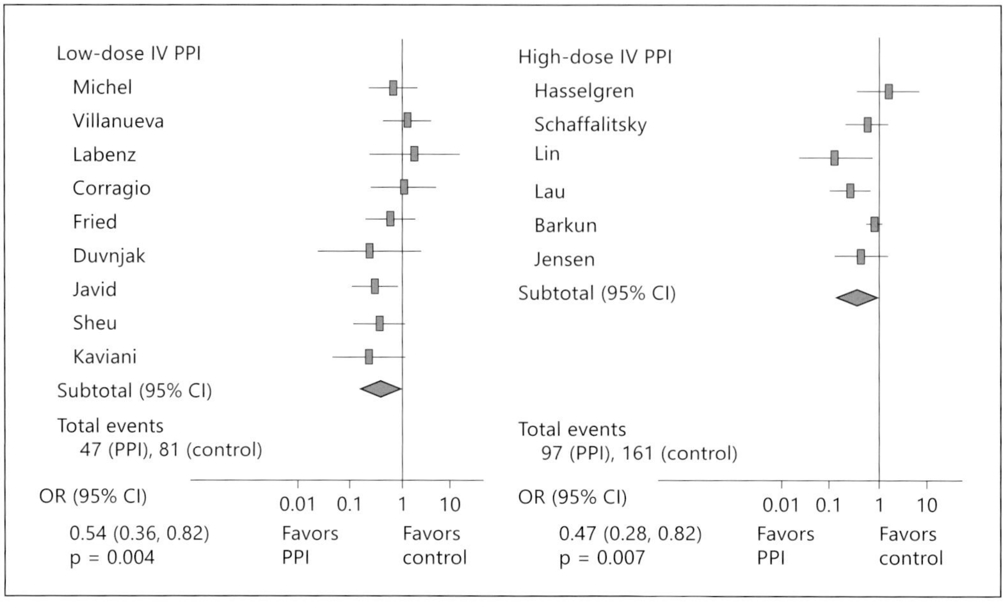

Fig. 3. PPIs (low- or high-dose) after endoscopic hemostasis in preventing recurrent bleeding from peptic ulcer disease [for details of the studies see 15].

72 h, or an intensive regimen, consisting of a 80-mg PPI bolus at the outset followed by a continuous infusion of 8 mg/h for 72 h, the clinical outcome was quite different [17]. Those who received the intensive regimen had a significantly lower recurrent bleeding rate (6.4%) compared to those who received the low-dose infection (11%).

The rationale to doubt whether low-dose PPIs can reduce recurrent bleeding is that in order to regain platelet aggregation and hence clot formation, the acidity of environment should be kept at above pH 6. Intermittent injections and low-dose intravenous infusion may not sustain a total 'shut down' of gastric acid secretion and hence a close-to-neutral pH in the stomach.

Would High-Dose Oral Proton Pump Inhibitors Work?

Interestingly, there are at least three randomized trials showing that oral PPI, when given with or without endoscopic therapy, may be able to reduce the risk of recurrent bleeding from peptic ulcer disease [18–20]. Incidentally, all three studies comparing oral omeprazole versus placebo come from South Asia, and the use of endoscopic therapy is not standardized. It is not clear why intermittent doses of intravenous PPIs may not work while oral PPIs seem to produce the beneficial effect.

Two studies have compared intragastric pH after administering oral and intravenous PPIs. In the first study, patients with overt ulcer bleeding received intravenous

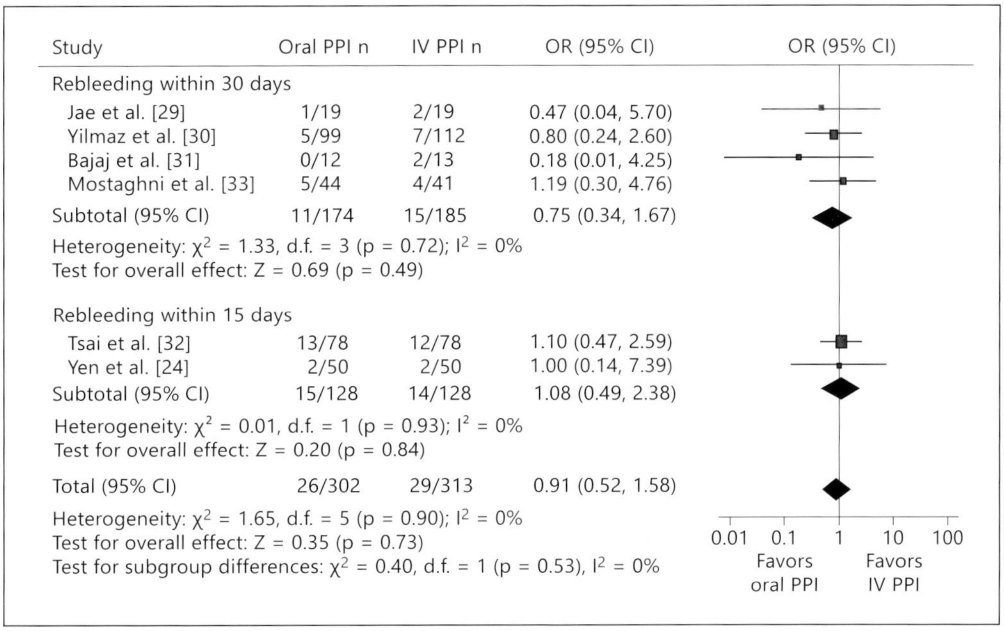

Fig. 4. Comparing oral vs. intravenous PPIs in preventing recurrent peptic ulcer bleeding [for details of the studies see 24, 29–33].

Table 2. Use of oral PPI in prevention of recurrent bleeding from peptic ulcer disease

Study	Endo Rx	Drug/regimen	n	Re-bleeding	p	Comments
Khuroo et al. [18]	none	omeprazole 40 mg q12h × 5 days placebo	110 110	10.9% 36.4%	<0.001	>benefit VV, clots ARR = 25.5 NNT = 3
Javid et al. [19]	yes	omeprazole 40 mg q12h × 5 days placebo	82 84	7.0% 21.0%	0.02	ARR = 14 NNT = 7
Kaviani et al. [20]	yes	omeprazole 40 mg q12h × 5 days placebo	71 78	17.0% 33.0%	0.022	ARR = 16 NNT = 6

VV = Visible vessel; ARR = absolute risk reduction; NNT = number needed to treat.

lansoprazole (90-mg bolus followed by 9-mg/h infusion) or oral lansoprazole (120-mg bolus followed by 30 mg every 3 h). Intragastric pH was higher than 6 for 67.8% of the study period with intravenous PPI and 64.8% with oral PPI (difference: 3 percentage points, 95% CI: –9.2 to 15.2) [21]. In the second study, a variety of oral and intravenous PPIs were used [22]. The mean 72-hour intragastric pH for oral omeprazole (80-mg bolus followed by 40 mg every 12 h) was 6.56 versus 6.93 for omeprazole infusion (80-mg bolus followed by 8 mg/h). The mean 72-hour intragastric pH for oral pantoprazole (80-mg bolus followed by 80 mg every 12 h) was 6.34 versus 6.43 for pantoprazole infusion (80-mg bolus followed by 8 mg/h). The mean 72-hour intragastric pH for oral rabeprazole (80-mg bolus followed by 40 mg every 12 h) was 6.11 versus 6.18 for rabeprazole infusion (80-mg bolus followed by 8 mg/h). This is significantly different from patients who had received no PPI for which the mean 72-hour pH was 2.04.

A recent study from Hong Kong compared high-dose oral PPIs versus high-dose intravenous PPIs in preventing recurrent peptic ulcer bleeding [23]. The study was double-blinded with a double-dummy design, and patients received either oral esomeprazole 40 mg twice daily or an intravenous esomeprazole 80-mg bolus injection followed by an 8-mg/h infusion for 72 h. Two hundred and fifteen patients with Forrest I or IIa/b peptic ulcer were treated with endoscopic therapy followed by PPIs. The results of the study showed that there was no difference in recurrent bleeding, hospital stay, blood transfusion, and surgery between the two groups. A small study from Taiwan giving patients either oral lansoprazole or intravenous esomeprazole also showed no difference in clinical outcome [24]. The comparable efficacy of oral versus intravenous PPIs was also confirmed by a pooled data analysis from a recent meta-analysis [25]. Therefore, in resource-limited practices, oral PPIs (at least 80 mg/day) can be given to patients after endoscopic therapy when oral feeding is resumed. The dose can be reduced to a regular level after recurrent bleeding has stopped for 3 days.

Should Proton Pump Inhibitors Be Given before Endoscopy?

In clinical practice, treatment with PPIs is often initiated before endoscopy in patients presenting with upper gastrointestinal bleeding. However, there is a lack of evidence in the literature to provide support for such a preemptive approach. A randomized controlled study was designed to test if early intravenous infusion of a high-dose PPI before endoscopy would have a therapeutic effect on bleeding ulcers, reduce the need for endoscopic therapy, and result in improved clinical outcomes [26]. Patients were randomly assigned to receive an intravenous infusion of omeprazole or placebo, each given as an 80-mg intravenous bolus injection followed by continuous infusion of 8 mg/h until endoscopic examination the next morning. The result showed that preemptive treatment with an intravenous PPI mitigates the risk of active bleeding during endoscopy and hence reduces the need for therapeutic endoscopy. However, the pre-endoscopy use of PPIs does not change the clinical outcome of peptic ulcer bleeding. The requirement for surgery and blood transfusions, as well as mortality, were not significantly changed. Therefore, the preemptive use of PPIs should not be routinely used. It is recommended when emergency endoscopy is not available throughout the day, and it should be used as a stop-gap therapy before endoscopy can be offered [27, 28].

References

1 Green FW Jr, Kaplan MM, Curtis LE, Levine PH: Effect of acid and pepsin on blood coagulation and platelet aggregation. A possible contributor prolonged gastroduodenal mucosal hemorrhage. Gastroenterology 1978;74:38–43.

2 Chaimoff C, Creter D, Djaldetti M: The effect of pH on platelet and coagulation factor activities. Am J Surg 1978;136:257–259.

3 Patchett SE, Enright H, Afdhal N, O'Connell W, O'Donoghue DP: Clot lysis by gastric juice: an in vitro study. Gut 1989;30:1704–1707.

4 Hasselgren G, Lind T, Lundell L, Aadland E, Efskind P, Falk A, et al: Continuous intravenous infusion of omeprazole in elderly patients with peptic ulcer bleeding. Results of a placebo-controlled multicenter study. Scand J Gastroenterol 1997;32:328–333.

5 Schaffalitzky de Muckadell OB, Havelund T, Harling H, Boesby S, Snel P, Vreeburg EM, et al: Effect of omeprazole on the outcome of endoscopically treated bleeding peptic ulcers. Randomized double-blind placebo-controlled multicentre study. Scand J Gastroenterol 1997;32:320–327.

6 Jensen DM, Pace SC, Soffer E, Comer GM, 315 Study Group: Continuous infusion of pantoprazole versus ranitidine for prevention of ulcer rebleeding: a U.S. multicenter randomized, double-blind study. Am J Gastroenterol 2006;101:1991–1999, quiz 2170.

7 Lau JY, Sung JJ, Lee KK, Yung MY, Wong SK, Wu JC, et al: Effect of intravenous omeprazole on recurrent bleeding after endoscopic treatment of bleeding peptic ulcers. N Engl J Med 2000;343:310–316.

8 Lin HJ, Lo WC, Cheng YC, Perng CL: Role of intravenous omeprazole in patients with high-risk peptic ulcer bleeding after successful endoscopic epinephrine injection: a prospective randomized comparative trial. Am J Gastroenterol 2006;101:500–505.

9 Sung JJ, Barkun A, Kuipers EJ, et al: Intravenous esomeprazole for prevention of recurrent peptic ulcer bleeding: a prospective randomized trial. Ann Intern Med 2009;150:455–464.

10 Lanas A, Polo-Tomas M, García-Rodríguez LA, et al: Effect of proton pump inhibitors on the outcomes of peptic ulcer bleeding: comparison of event rates in routine clinical practice and a clinical trial. Scand J Gastroenterol 2013;48:285–294.

11 Daneshmend TK, Hawkey CJ, Langman MJ, Logan RF, Long RG, Walt RP: Omeprazole versus placebo for acute upper gastrointestinal bleeding: randomised double blind controlled trial. BMJ 1992;304:143–147.

12 Ortí E, Canelles P, Quiles F, Zapater R, Cuquerella J, Ariete V, Tomé A, Medina E: Does the antisecretory agent used affect the evolution of upper digestive hemorrhage (in Spanish)? Rev Esp Enferm Dig 1995; 87:427–430.

13 Villanueva C, Balanzó J, Torras X, Sáinz S, Soriano G, González D, Vilardell F: Omeprazole versus ranitidine as adjunct therapy to endoscopic injection in actively bleeding ulcers: a prospective and randomized study. Endoscopy 1995;27:308–312.

14 Lanas A, Artal A, Blás JM, Arroyo MT, Lopez-Zaborras J, Sáinz R: Effect of parenteral omeprazole and ranitidine on gastric pH and the outcome of bleeding peptic ulcer. J Clin Gastroenterol 1995;21:103–106.

15 Leontiadis GI, Sharma VK, Howden CW: Proton pump inhibitor treatment for acute peptic ulcer bleeding. Cochrane Library 2006;1:CD002094.

16 Hsu YC, Perng CL, Yang TH, Wang CS, Hsu WL, Wu HT, Cheng YC, Chiang MF, Lin HJ: A randomized controlled trial comparing two different dosages of infusional pantoprazole in peptic ulcer bleeding. Br J Clin Pharmacol 2010;69:245–251.

17 Liu N, Liu L, Zhang H, Gyawali PC, Zhang D, Yao L, Yang Y, Wu K, Ding J, Fan D: Effect of intravenous proton pump inhibitor regimens and timing of endoscopy on clinical outcomes of peptic ulcer bleeding. J Gastroenterol Hepatol 2012;27:1473–1479.

18 Khuroo MS, Yattoo GN, Javid G, Khan BA, Shah AA, Gulzar GM, et al: A comparison of omeprazole and placebo for bleeding peptic ulcer. N Engl J Med 1997; 336:1054–1058.

19 Javid G, Masoodi I, Zargar SA, Khan BA, Yatoo GN, Shah AH, Gulzar GM, Sodhi JS: Omeprazole as adjuvant therapy to endoscopic combination injection sclerotherapy for treating bleeding peptic ulcer. Am J Med 2001;111:280–284.

20 Kaviani MJ, Hashemi MR, Kazemifar AR, Roozitalab S, Mostaghni AA, Merat S, Alizadeh-Naini M, Yarmohammadi H: Effect of oral omeprazole in reducing re-bleeding in bleeding peptic ulcers: a prospective, double-blind, randomized, clinical trial. Aliment Pharmacol Ther 2003;17:211–216.

21 Laine L, Shah A, Bemanian S: Intragastric pH with oral vs intravenous bolus plus infusion proton-pump inhibitor therapy in patients with bleeding ulcers. Gastroenterology 2008;134:1836–1841.

22 Javid G, Zargar SA, U-Saif R, et al: Comparison of p.o. or i.v. proton pump inhibitors on 72-h intragastric pH in bleeding peptic ulcer. J Gastroenterol Hepatol 2009;24:1236–1243.

23 Sung JJ, Suen BY, Wu JCY, et al: Effects of intravenous and oral esomeprazole in prevention of recurrent bleeding from peptic ulcers after endoscopic therapy (abstract). Gastroenterology 2012;142.5: S-192.

24 Yen HH, Yang CW, Su WW, et al: Oral versus intravenous proton pump inhibitors in preventing rebleeding for patients with peptic ulcer bleeding after successful endoscopic therapy. BMC Gastroenterol 2012;12:66.

25 Tsoi KW, Hirai H, Sung JJY: Oral and intravenous proton pump inhibitors for patients with peptic ulcer bleeding: a meta-analysis of prospective randomized trials (abstract). Gut 2012;61:A7.
26 Lau JY, Leung WK, Wu JC, Chan FK, Wong VW, Chiu PW, Lee VW, Lee KK, Cheung FK, Siu P, Ng EK, Sung JJ: Omeprazole before endoscopy in patients with gastrointestinal bleeding. N Engl J Med 2007;356:1631–1640.
27 Barkun AN, Bardou M, Kuipers EJ, Sung J, Hunt RH, Martel M, Sinclair P, International Consensus Upper Gastrointestinal Bleeding Conference Group: International consensus recommendations on the management of patients with nonvariceal upper gastrointestinal bleeding. Ann Intern Med 2010;152:101–113.
28 Sung JJ, Chan FK, Chen M, Ching JY, Ho KY, Kachintorn U, Kim N, Lau JY, Menon J, Rani AA, Reddy N, Sollano J, Sugano K, Tsoi KK, Wu CY, Yeomans N, Vakil N, Goh KL, Asia-Pacific Working Group: Asia-Pacific Working Group consensus on non-variceal upper gastrointestinal bleeding. Gut 2011;60:1170–1177.
29 Jae YJ, Kwang RJ, Young H, et al: A comparison of the effect of high-dose oral and intravenous proton pump inhibitor on the prevention of rebleeding after endoscopic treatment of bleeding peptic ulcers. Korean J Gastrointest Endosc 2006;22:6–11.
30 Yilmaz S, Bayan K, Tuzan Y, et al: A head to head comparison of oral vs intravenous omeprazole for patients with bleeding peptic ulcers with a clean base, flat spots and adherent clots. World J Gastroenterol 2006;12:7837–7843.
31 Bajaj JS, Dua KS, Hanson K, Presberg K: Prospective, randomized trial comparing effect of oral versus intravenous pantoprazole on rebleeding after nonvariceal upper gastrointestinal bleeding: a pilot study. Dig Dis Sci 2007;52:2190–2194.
32 Tsai JJ, Hsu YC, Perng CL, Lin HJ: Oral or intravenous proton pump inhibitor in patients with peptic ulcer bleeding after successful endoscopic epinephrine injection. Br J Clin Pharmacol 2008;67:326–332.
33 Mostaghi AA, Hashemi SA, Heydari ST: Comparison of oral and intravenous proton pump inhibitor on patients with high risk bleeding ulcers: a prospective, randomized, controlled clinical trial. Iran Red Crescent Med J 2011;13:458–463.

Joseph Sung, MD, PhD
VC Office, The Chinese University of Hong Kong
Shatin, NT, Hong Kong, SAR (China)
E-Mail joesung@cuhk.edu.hk

Chiba T, Malfertheiner P, Satoh H (eds): Proton Pump Inhibitors: A Balanced View.
Front Gastrointest Res. Basel, Karger, 2013, vol 32, pp 77–83 (DOI: 10.1159/000350633)

Proton Pump Inhibitors in Functional Dyspepsia

Hiroto Miwa

Division of Upper Gastroenterology, Hyogo College of Medicine, Nishinomiya, Japan

Abstract

Functional dyspepsia (FD) refers to the chronic presence of symptoms thought to originate in the gastroduodenal region in the absence of organic disease that explains the symptoms. The pathogenesis of FD is complex and multifactorial. FD symptoms are induced by stress, and differences in individual response to stress exist in the background. Acid is considered to be a factor that modifies the manifestation of FD symptoms. Although it has not been proven that excessive gastric acid is present in FD patients, proton pump inhibitors (PPIs) are widely used as first-line drugs in the treatment of FD. Meta-analyses of clinical studies of the efficacy of PPIs in FD have shown that they are effective but have limited efficacy, with treatment of 9–15 patients needed to cure 1 patient. PPIs may be effective in FD patients with undiagnosed gastroesophageal reflux disease.

Functional gastrointestinal disorder is a general term for gastrointestinal disorders in which symptoms are caused by functional abnormalities rather than by organic causes. Under the influence of the steadily growing complexity and associated stress of modern society, the incidence of functional gastrointestinal disorders such as irritable bowel syndrome (IBS) and functional dyspepsia (FD) is increasing. FD has been defined as 'the presence of symptoms thought to originate in the gastroduodenal region, in the absence of any organic, systemic, or metabolic disease that is likely to explain the symptoms' [1]. The diagnostic criteria for FD also require that such symptoms are chronic [1].

Although everyone occasionally experiences dyspeptic symptoms such as a stomachache or a heavy feeling in the stomach, FD is a highly prevalent, serious condition that can significantly impair a patient's quality of life and affect society by reducing labor productivity. FD is difficult to treat in general, and the efficacy of pharmacotherapy in FD is not great, but among the drugs used to treat FD, PPIs are a first-line option. This chapter reviews the role of acid in the pathogenesis of FD and discusses the efficacy of proton pump inhibitors (PPIs) as a treatment for FD.

Model of Functional Dyspepsia Pathogenesis

Intuitively, we tend to expect there to be some abnormality at the site where symptoms occur; however, since organic disease has been ruled out in FD, the patient's symptoms are attributed to abnormalities in gastroduodenal function that include abnormal gastric motility, visceral hypersensitivity, and abnormal secretion of gastric acid.

A problem in the diagnosis and treatment of functional gastrointestinal disorders is that the symptoms occur intermittently. The patients are prone to symptoms, but do not experience symptoms continuously. In IBS, some patients experience abdominal pain and other symptoms, along with defecation abnormalities. In a recent large-scale internet survey of IBS patients [2], patients with diarrhea-type IBS that fulfilled the Rome III criteria were questioned about their frequency of defecation. During asymptomatic periods, the average frequency was 1.4 times/day, which did not differ significantly from the frequency in healthy persons (1.5 times/day); however, during symptomatic periods, the frequency increased to 3.5 times/day. In other words, the IBS patients normally did not differ from healthy persons, but sometimes they experienced symptoms caused by abnormal functioning of the digestive tract. In addition, the symptoms were triggered by stress. Fukudo et al. [3] intravenously administered corticotropin-releasing hormone, which is secreted under stress, to IBS patients and healthy controls and recorded colonic motility. Corticotropin-releasing hormone increased colonic motility significantly in both the IBS patients and controls, but the increases were much greater in the patients. This result means that the digestive tract of patients with functional gastrointestinal disorders responds excessively to stress.

Considering the evidence related to the pathogenesis of FD that has been collected to date, I think FD symptoms can be explained by a model in which patients who for some reason have acquired excessive responsiveness to stress directly manifest symptoms due to abnormalities in physiological function in the upper gastrointestinal tract, with that process modified by a variety of factors [4]. According to this model, the factors related to the manifestation of FD symptoms can be divided into three categories – factors that are directly related to symptom manifestation, factors that indirectly modify symptom manifestation, and factors that govern responsiveness to stress. However, the essence of FD pathogenesis may be the factors that govern the responsiveness of FD patients to stress, including environment during early life, genetic abnormalities, and residual inflammation after gastrointestinal infections. The abnormalities in physiological function in the upper gastrointestinal tract that directly result in symptoms are abnormal motility and hypersensitivity. Acid (like *Helicobacter pylori* infection, psychological factors, diet, and lifestyle) is thought to be a factor that modifies abnormal motility and hypersensitivity.

Role of Acid in Functional Dyspepsia Pathophysiology

According to the model described above, the role of acid in FD pathogenesis is to indirectly modify the manifestation of FD symptoms. In other words, acid is thought to be an indirect factor. However, the model has not been widely accepted, and the idea that acid secretion is one of the most important physiological functions of the stomach and that acid is an important direct cause of FD has deep roots. Indeed, dyspepsia symptoms occur when duodenal mucosa are directly exposed to acid [5], and postprandial duodenal pH is lower in FD patients than in healthy controls [6]. Furthermore, pathological acid reflux into the esophagus is present even in a subset of FD patients without heartburn [7]. However, experimental exposure of the duodenum or esophagus to acid causes functional abnormalities including gastrointestinal hypersensitivity and abnormal gastroduodenal motility [8–12]. Therefore, it can be inferred that induction of FD symptoms by acid is mediated by such hypersensitivity and abnormal motility.

To investigate the relationship between gastric acid and dyspepsia symptoms, we introduced acid and water into the stomach of healthy volunteers, and found that dyspepsia symptoms were significantly more severe after the introduction of acid [13]. This result suggested that acid caused the dyspepsia symptoms. However, when we subsequently performed a similar experiment with healthy volunteers and FD patients, the symptoms were significantly greater in the FD patients, even though the amount of acid introduced in the stomach was the same in the FD patients and healthy subjects (fig. 1) [14]. From these results, we inferred that the difference between FD patients and healthy persons may be explained by the responsiveness of the body to acid rather than by the amount of acid. If this is the case, it explains why healthy persons do not have dyspepsia symptoms even though they secrete gastric acid.

Role and Efficacy of Proton Pump Inhibitors as Pharmacotherapeutic Agents for Functional Dyspepsia

If the essence of the pathogenesis of FD is an excessive response of the digestive system to stress, psychosomatic approaches that focus on stress management should be effective. Indeed, cognitive-behavioral therapy [15, 16], hypnotherapy [17], and other psychosomatic therapies have been found to be effective; however, use of such therapies requires special training and techniques, and thus many clinicians are not comfortable using such therapies. Since most FD patients are cared for by primary care

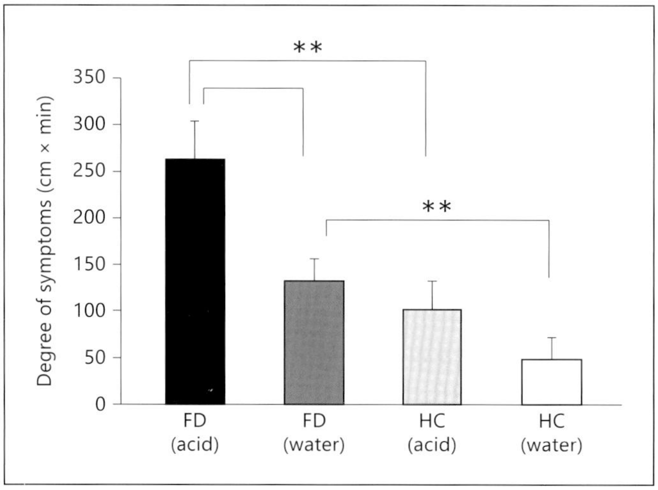

Fig. 1. Comparison of the response to acid by FD patients and healthy controls. Either 0.1 mol/l of hydrochloric acid or water was instilled in the stomach of FD patients (FD) and healthy controls (HC) in a double-blind fashion, and the degree of symptoms was determined from symptom intensity and duration. Acid provoked greater symptoms than water in both the FD patients and the healthy controls, but the FD patients had a greater response to both water and acid than the controls. These results suggest that the difference between FD patients and healthy persons may be determined by hypersensitivity to gastric acid rather than by amount of acid. ** $p < 0.01$. Adapted from [14].

physicians, FD is usually treated with drugs. Unfortunately, however, the efficacy of these drugs is not high. One reason for this is the extremely high placebo response rates seen in FD patients [18]. There have been almost no studies in which the response of FD patients to a drug has greatly exceeded the response to placebo, and for this reason, one cannot expect PPIs to have much efficacy in FD patients. In addition, the theory that PPIs should be an effective treatment for FD is also not persuasive from the viewpoint of the mechanism of PPI action since it has not been proven that excessive gastric acid is present in FD patients.

Nevertheless, acid secretion inhibitors and prokinetics are commonly used as first-line drugs for treatment of FD [19–21]. The reasons for this include the following: (1) although the efficacy of acid secretion inhibitors and prokinetics is not high, it is supported by reliable evidence; (2) acid is easy to set as a therapeutic target, and PPIs are widely used; (3) PPIs are highly effective in patients with uninvestigated dyspepsia, and (4) there is considerable overlap of patients with nonerosive reflux disease (NERD) and patients with FD.

Because dyspepsia is an extremely common condition, physicians often treat patients with dyspepsia symptoms on the basis of experience without confirming the diagnosis by endoscopy or other definitive diagnostic modality. Such patients are regarded as having uninvestigated dyspepsia, and in them, PPIs are highly effective. PPIs

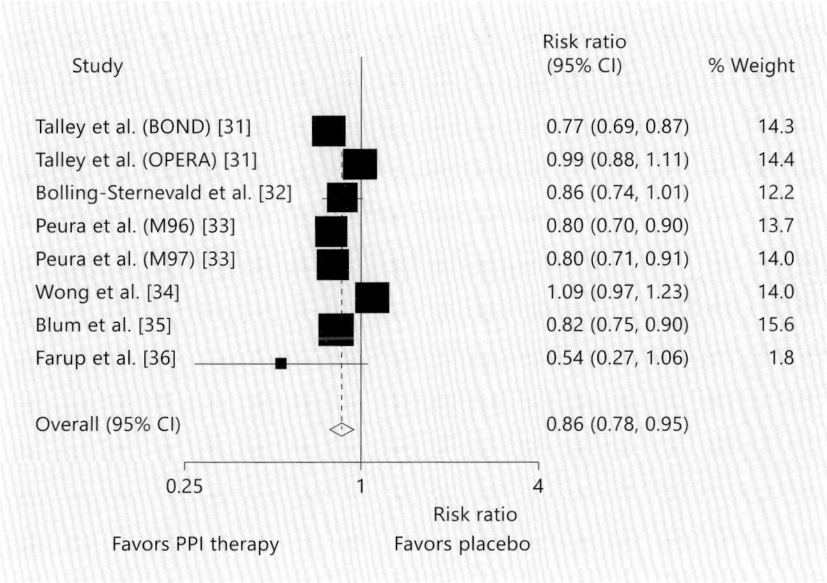

Fig. 2. Meta-analysis of the efficacy of PPIs in FD. Results of 8 randomized, controlled trials of PPI therapy vs. placebo in patients with FD. Overall results demonstrated significantly positive efficacy of PPIs (OR: 0.96; 95% CI: 78–95%) even though 4 of the 8 trials did not show any significant difference. The number needed to treat for PPIs from this meta-analysis was approximately 9. Adapted from [27].

have been especially effective in uninvestigated dyspepsia in populations with low *H. pylori* infection rates [22, 23]. This is inferred to be because patients with gastroesophageal reflux disease (GERD) account for the greatest proportion of patients with uninvestigated dyspepsia. Another hypothesis to explain the efficacy of PPIs in dyspepsia is overlap of NERD patients and FD patients [24–26], which is known to occur at a high frequency. Even if NERD is not diagnosed directly from heartburn symptoms, pathological esophageal acid exposure is not uncommon in FD patients [7].

How effective are PPIs in FD patients in whom organic disease has been ruled out by endoscopy or another modality, and heartburn is not the chief complaint? This question has been investigated in a number of comparatively high-quality clinical studies. A systematic review to evaluate whether PPI therapy has any effect in nonulcer dyspepsia that was published in 2004 and included eight studies found that PPIs had a significant ameliorative effect on FD symptoms, but the relative risk of remaining dyspeptic with PPI therapy versus placebo was 0.86 (95% CI: 0.78–0.95) and the number needed to treat to cure 1 FD patient was 9 [27] (fig. 2). A meta-analysis of seven studies published by a Hong Kong group in 2007 also found that PPIs were significantly more effective than placebo for reducing symptoms in patients with FD, but the relative risk reduction was only 0.10 (95% CI: 0.03–0.17) and the number needed to treat to cure 1 FD patient was 14.6 [28]. After those meta-analyses, a clinical study of the efficacy of the PPI pantoprazole in patients with ulcer-like dyspepsia found that pantoprazole had a significant symptom-relieving effect, but the effect was not especially large [29]. PPIs have been reported to be effective in FD patients with pain as the chief complaint [28–30]; therefore, such patients may be good candidates for PPI therapy. However, based on the results of clinical studies of PPIs, it can probably be said that while PPIs certainly are effective, generally their effect is not large. Furthermore, the finding that differences in relief of FD symptoms are not seen between low-dose and normal-dose PPI therapy suggests that the efficacy of PPIs in FD is limited [28].

References

1 Tack J, Talley NJ, Camilleri M, et al: Functional gastroduodenal disorders. Gastroenterology 2006;130:1466–1479.
2 Miwa H: Clinical feature of IBS patients in Japan – from large scaled Internet survey (in Japanese). Shinyaku Rinsyo 2011;60:2130–2147.
3 Fukudo S, Nomura T, Hongo M: Impact of corticotropin-releasing hormone on gastrointestinal motility and adrenocorticotropic hormone in normal controls and patients with irritable bowel syndrome. Gut 1998;42:845–849.
4 Miwa H: Why dyspepsia can occur without organic disease: pathogenesis and management of functional dyspepsia. J Gastroenterol 2012;47:862–871.
5 Ishii M, Manabe N, Kusunoki H, et al: Real-time evaluation of dyspeptic symptoms and gastric motility induced by duodenal acidification using noninvasive transnasal endoscopy. J Gastroenterol 2008;43:935–941.
6 Lee KJ, Demarchi B, Demedts I, et al: A pilot study on duodenal acid exposure and its relationship to symptoms in functional dyspepsia with prominent nausea. Am J Gastroenterol 2004;99:1765–1773.
7 Tack J, Caenepeel P, Arts J, et al: Prevalence of acid reflux in functional dyspepsia and its association with symptom profile. Gut 2005;54:1370–1376.
8 di Stefano M, Vos R, Vanuytsel T, et al: Prolonged duodenal acid perfusion and dyspeptic symptom occurrence in healthy volunteers. Neurogastroenterol Motil 2009;21:712–e40.
9 Hobson AR, Khan RW, Sarkar S, et al: Development of esophageal hypersensitivity following experimental duodenal acidification. Am J Gastroenterol 2004;99:813–820.
10 Lee KJ, Kim JH, Cho SW: Dyspeptic symptoms associated with hypersensitivity to gastric distension induced by duodenal acidification. J Gastroenterol Hepatol 2006;21:515–520.
11 Lee KJ, Vos R, Janssens J, et al: Influence of duodenal acidification on the sensorimotor function of the proximal stomach in humans. Am J Physiol Gastrointest Liver Physiol 2004;286:G278–G284.
12 van den Elzen BD, Tytgat GN, Boeckxstaens GE: Gastric hypersensitivity induced by oesophageal acid infusion in healthy volunteers. Neurogastroenterol Motil 2009;21:160–169.
13 Miwa H, Nakajima K, Yamaguchi K, et al: Generation of dyspeptic symptoms by direct acid infusion into the stomach of healthy Japanese subjects. Aliment Pharmacol Ther 2007;26:257–264.
14 Oshima T, Okugawa T, Tomita T, et al: Generation of dyspeptic symptoms by direct acid and water infusion into the stomachs of functional dyspepsia patients and healthy subjects. Aliment Pharmacol Ther 2012;35:175–182.
15 Haag S, Senf W, Tagay S, et al: Is there a benefit from intensified medical and psychological interventions in patients with functional dyspepsia not responding to conventional therapy? Aliment Pharmacol Ther 2007;25:973–986.
16 Haug TT, Wilhelmsen I, Svebak S, et al: Psychotherapy in functional dyspepsia. J Psychosom Res 1994;38:735–744.
17 Calvert EL, Houghton LA, Cooper P, et al: Long-term improvement in functional dyspepsia using hypnotherapy. Gastroenterology 2002;123:1778–1785.
18 Talley NJ, Locke GR, Lahr BD, et al: Predictors of the placebo response in functional dyspepsia. Aliment Pharmacol Ther 2006;23:923–936.
19 Tack J, Talley N: Gastroduodenal disorders. Am J Gastroenterol 2010;105:757–763.
20 Miwa H, Ghoshal UC, Fock KM, et al: Asian consensus report on functional dyspepsia. J Gastroenterol Hepatol 2012;27:626–641.
21 Talley NJ, Vakil N, Practice Parameters Committee of the American College of Gastroenterology: Guidelines for the management of dyspepsia. Am J Gastroenterol 2005;100:2324–2337.

22 Veldhuyzen van Zanten SJ, Chiba N, Armstrong D, et al: A randomized trial comparing omeprazole, ranitidine, cisapride, or placebo in *Helicobacter pylori* negative, primary care patients with dyspepsia: the CADET-HN Study. Am J Gastroenterol 2005;100: 1477–1488.
23 Sakurai K, Nagahara A, Inoue K, et al: Efficacy of omeprazole, famotidine, mosapride and teprenone in patients with upper gastrointestinal symptoms: an omeprazole-controlled randomized study (J-FOCUS). BMC Gastroenterol 2012;12:42.
24 Mahadeva S, Goh KL: Epidemiology of functional dyspepsia: a global perspective. World J Gastroenterol 2006;12:2661–2666.
25 Spechler SJ: Epidemiology and natural history of gastro-oesophageal reflux disease. Digestion 1992; 51(suppl 1):24–29.
26 Locke GR 3rd, Talley NJ, Fett SL, Zinsmeister AR, Melton LJ 3rd: Prevalence and clinical spectrum of gastroesophageal reflux: a population-based study in Olmsted County, Minnesota. Gastroenterology 1997; 112:1448–1456.
27 Moayyedi P, Delaney BC, Vakil N, et al: The efficacy of proton pump inhibitors in nonulcer dyspepsia: a systematic review and economic analysis. Gastroenterology 2004;127:1329–1337.
28 Wang WH, Huang JQ, Zheng GF, et al: Effects of proton-pump inhibitors on functional dyspepsia: a meta-analysis of randomized placebo-controlled trials. Clin Gastroenterol Hepatol 2007;5:178–185.
29 van Rensburg C, Berghöfer P, Enns R, et al: Efficacy and safety of pantoprazole 20 mg once daily treatment in patients with ulcer-like functional dyspepsia. Curr Med Res Opin 2008;24:2009–2018.
30 Talley NJ, Lauritsen K: The potential role of acid suppression in functional dyspepsia: the BOND, OPERA, PILOT, and ENCORE studies. Gut 2002; 50(suppl 4):iv36–iv41.
31 Talley NJ, Meineche-Schmidt V, Pare P, et al: Efficacy of omeprazole in functional dyspepsia: double-blind, randomized, placebo-controlled trials (the BOND and OPERA studies). Aliment Pharmacol Ther 1998;12:1055–1065.
32 Bolling-Sternevald E, Lauritsen K, Aalykke C, et al: Effect of profound acid suppression in functional dyspepsia: a double-blind, randomized, placebo-controlled trial. Scand J Gastroenterol 2002;37:1395–1402.
33 Peura DA, Kovacs TO, Metz DC, et al: Lansoprazole in the treatment of functional dyspepsia: two double-blind, randomized, placebo-controlled trials. Am J Med 2004;116:740–748.
34 Wong WM, Wong BC, Hung WK, et al: Double blind, randomized, placebo controlled study of four weeks of lansoprazole for the treatment of functional dyspepsia in Chinese patients. Gut 2002;51:502–506.
35 Blum AL, Arnold R, Stolte M, et al: Short course acid suppressive treatment for patients with functional dyspepsia: results depend on *Helicobacteria pylori* status. The Frosch Study Group. Gut 2000;47:473–480.
36 Farup PG, Hovde O, Torp R, Wetterhus S: Patients with functional dyspepsia responding to omeprazole have a characteristic gastro-oesophageal reflux pattern. Scand J Gastroenterol 1999;34:575–579.

Prof. Hiroto Miwa, MD, PhD
Division of Upper Gastroenterology, Department of Internal Medicine
Hyogo College of Medicine, 1-1 Mukogawa-cho, Nishinomiya
Hyogo 663-8501 (Japan)
E-Mail miwahgi@hyo-med.ac.jp

Chiba T, Malfertheiner P, Satoh H (eds): Proton Pump Inhibitors: A Balanced View.
Front Gastrointest Res. Basel, Karger, 2013, vol 32, pp 84–91 (DOI: 10.1159/000350634)

Role of Acid Suppression in Exocrine Pancreatic Insufficiency: A Therapeutic Principle

A. Waldthaler · K. Schütte · P. Malfertheiner

Department of Gastroenterology, Hepatology and Infectious Diseases, Otto von Guericke University, Magdeburg, Germany

Abstract

The use of acid suppression in patients with exocrine pancreatic insufficiency (EPI) additionally to enteric-coated pancreatic enzyme preparations (PEPs) is controversial. In current recommendations, the prescription of proton pump inhibitors (PPIs) is not generally recommended, but mentioned as an auxiliary agent if PEPs alone are not effective and a dose augmentation does not lead to a satisfying improvement of symptoms. Adverse effects of PPI treatment are rare while several beneficial aspects of their use in patients with EPI have become evident. The addition of PPIs to PEPs in patients with EPI results in a higher duodenal pH and thus a reduced inactivation of the residual enzyme secretion of patients, an earlier dissolution of PEPs from enteric coating, and a prevention of bile salt precipitation. These factors contribute to a more effective and physiological digestive pattern in patients with EPI. This article gives a detailed overview of why we believe the recommendation for the use of acid-suppressing drugs in patients with EPI is justified.

The pancreas fulfills exocrine and endocrine functions. The cells with endocrine functions are located in the islets of Langerhans and produce insulin, glucagon, somatostatin, ghrelin, and pancreatic polypeptide. The exocrine part of the gland is composed of acinar cells and secretes enzymes and bicarbonate.

When chymus enters the duodenum, pancreatic secretion is stimulated by the release and synergistic interaction of secretin and cholecystokinin [1]. These hormones exert a negative feedback on gastrin and thereby on gastric acid production and motility of the stomach and duodenum [2].

Together with the pancreatic secretion of bicarbonate, this results in a neutralization of gastric acid and duodenal alkalization. This mechanism is crucial for pancreatic enzymes to work optimally [3]. Pancreatic acinar cells secrete carbohydrases, proteases, and lipase for the digestion of carbohydrates, protein, and fat. Lipase (gastric lipase, intestinal lipase), carbohydrases, and other proteases are also produced in other regions of the upper gastrointestinal tract, but especially for lipase, the amounts outside the pancreas are of minor relevance for digestion.

Effects of Chronic Pancreatitis on Digestion

Chronic pancreatitis (CP) is an inflammatory disease predominantly caused by chronic alcohol abuse and less frequently by hereditary conditions and other rare causes. CP is characterized by progressive destruction of the pancreatic tissue, which leads to abdominal pain and loss of endocrine and exocrine function. The time of onset of maldigestion due to exocrine pancreatic insufficiency (EPI) is not predictable, but the probability increases with duration of disease [4] and the first impairment in exocrine function is a reduction of pancreatic bicarbonate output [5]. In alcoholic CP, exocrine insufficiency is present in 50% of patients after 10 years, while after 20 years from onset of CP almost all patients suffer from it [6]. In nonalcoholic CP, the onset of exocrine insufficiency is usually later, but when present, it does not differ in characteristics from alcoholic CP [4, 7]. The most common inherited cause for EPI is cystic fibrosis (CF), and similar changes in digestion have been described in patients with EPI due to CF and CP [8–11]. Usually, the first symptoms of EPI appear after the gland has lost >90% of its enzyme output capacity, resulting in steatorrhea [12].

Malabsorption affects primarily fat-soluble nutritional components; the digestion of other nutrients is usually impaired only to a minor extent [13]. Several factors are responsible for this observation: the capacity of the pancreas to synthesize and secrete lipase is impaired first and more severely than other enzymes, and lipase is more susceptible to proteolytic destruction. Also proteins and carbohydrates are in part digested by enzymes released with saliva (amylase) and gastric juice (pepsin) [7, 12, 14, 15].

The most important part of physiological fat digestion and absorption takes place in the duodenum to the ligament of Treitz [3, 16, 17]. The late appearance of symptoms of EPI is partly related to a shift of digestion and absorption from the duodenum to more distal parts of the small bowel [3] (fig. 1). This might be a reason why small intestine bacterial overgrowth is very common in patients with CP [18]. Also changes in the activity of the intestinal brush border enzymes disaccharidase and alkaline phosphatase have been described in patients with EPI due to CP [19].

Increased delivery of nutrients to the more distal parts of the small bowel may result in disturbed regulation of motor and secretory functions of the upper gastrointestinal tract, but data on changes in gastrointestinal motility in CP are complex and heterogeneous, ranging from hyper- to hypomotility [16, 20, 21]. This might in part be explained by different study designs, but also by differences in comorbidity and changes of motility during the course of disease, by alcoholic and diabetic neuropathy or small intestine bacterial overgrowth. Another factor contributing to malnutrition is abdominal pain, especially when related to meals, which leads to a reduction in oral food intake with avoidance of fatty meals [22].

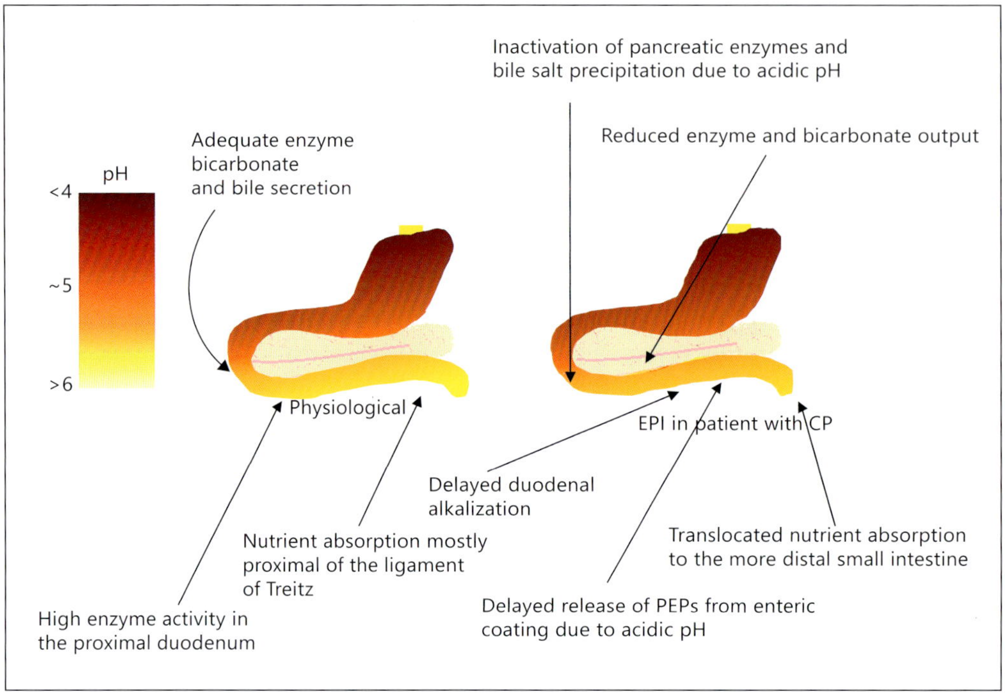

Fig. 1. Digestive pattern shifted to more distal parts of the small intestive in pancreatic insufficiency (EPI) due to chronic pancreatitis. PEPs = Pancreatic enzyme preparations.

Acid in Chronic Pancreatitis

In EPI, not only the enzyme output of acinar cells is impaired, but there is also a close correlation between pancreatic enzyme output and bicarbonate secretion capacity from pancreatic ductal epithelial cells [23]. At the same time, gastric acid production is increased [24, 25] and negative feedback to gastric acid production is insufficient in patients with CP [26]. As a consequence, duodenal and gastric pH are lower in patients with EPI due to CP and may fall under pH 4 postprandially [23, 25, 27, 28].

Acidic duodenal pH leads to an irreversible inactivation of pancreatic enzymes, especially of lipase, and only a part of the enzymes secreted become available for digestion [29, 30]. The decrease of duodenal pH also leads to a precipitation of bile salts and secondary bile salt malabsorption [3, 8, 30, 31].

Pharmacological Treatment of Exocrine Pancreatic Insufficiency

Oral substitution of pancreatic enzyme preparations (PEPs) is the obvious therapeutic approach for EPI. The general requirements for PEPs are high lipase activity, resistance to gastric acid, homogenous mixing and contemporaneous passage with chymus, and early activation in the duodenum. Different compositions and galenic forms have been developed and tested over the years with variable results [32–35]. When comparing study data, the evolution of galenics, especially concerning particle size and coating, has to be considered.

Today, commercially available PEPs are mostly of porcine origin and differ in their composition, dissolution profile, and dosage form. To protect enzymes from destruction by gastric acid, they are coated with a protective film called an enteric coating, which dissolves when the acid is neutralized, between pH 5 and 6, corresponding to the duodenal pH in healthy individuals [36, 37].

The standard administration form is a rapidly dissolving capsule that contains 100–800 particles of PEPs called micropellets or microspheres [37, 38]. Modern preparations contain microspheres of 1–3 mm in diameter with a coating on every single microsphere to guarantee the best possible mixing of chymus and enzymes, and simultaneous passage into the duodenum [37, 38].

The choice of PEP seems to influence the individual therapeutic effect, and there seem to be major differences between PEPs [37]. A change of PEP in case of therapeutic failure is advisable [39]. Enzymes should be ingested during a meal and recommended dosages for initial therapy depend on the amount of fat ingested.

The exact substitution dose recommendation for PEP is 2,000 IU of lipase per gram of fat ingested [40]. In most cases, doses per meal range from 25,000 to 40,000 units. In case of therapy failure, the first step in treatment recommendations is to increase the dosage [41, 42]. An overdose or intoxication with PEPs is not possible since the luminal enzymes are not absorbed, but a dose-dependent risk of fibrotic colon stenoses has been described in patients with CF using enteric-coated PEPs [43], with a typical location in the colon ascendens. This observation might be related to high active enzyme availability in the more distal parts of the small, and maybe also proximal large, bowel [43, 44].

The opinions on the effect of PEPs on pain in CP are divergent. A meta-analysis of the literature published on this topic showed no significant effect [45], but abnormal motility associated with CP showed an improvement after administration of PEP [16, 46], which might also lead to a reduction of postprandial dyspeptic symptoms. Although EPI is a well-known comorbidity in CP, most patients are undertreated – even in countries with well-organized health care systems [42].

The Role of Acid Suppression in Treatment of Exocrine Pancreatic Insufficiency

The use of acid suppression in addition to an enteric-coated PEP is controversial. The prescription of proton pump inhibitors (PPIs) is not generally recommended in the current recommendations for the use of PEPs, but it is mentioned as an auxiliary agent if a PEP alone is not effective and a dose augmentation of the PEP does not lead to a satisfying improvement of symptoms [41, 47].

The risks of PPI treatment are low, but a facilitation of bacterial overgrowth has been reported, but data are conflicting. CP is a predisposing factor for small bowel bacterial overgrowth as well [48, 49].

The use of PPIs leads to an increase in postprandial gastric and small intestine pH levels in healthy individuals as well as in patients with CP, correcting duodenal acidic pH [27, 50]. Although PPIs inhibit gastric acid secretion, they have no direct negative effect on the secretion of pancreatic acinar cells [51].

Gastric acid suppression either by the use of H_2-antagonists or PPIs in combination with an enteric-coated PEP significantly improves maldigestion in CP patients [52–54] and patients with CF [10, 11, 55–57] with EPI. Correction of duodenal acidic pH levels has resulted in a significant decline in fecal fat and excretion, bile salt precipitation, and improvement of nutritional status [10, 11, 52–57].

These observations might be explained by various factors. An elevation of duodenal pH avoids inactivation of the patient's residual enzyme output. A second aspect is that the release of PEPs from enteric coating is pH dependent, and delayed alkalization of the chymus also causes a delayed release of pancreatic enzymes from the enteric coating [34, 36]. Luminal release of enzymes from PEPs occurs only to a minor part in the duodenum, and a higher availability of enzymes in the ileum has been shown when using enteric-coated PEPs [29, 34]. In this way the digestive potential of lipase and the adsorptive capacity of the small intestine are not optimally used, which might be an explanation for persistent steatorrhea in many patients ingesting high doses of PEPs [29]. Conversely, a reduction of PEPs in combination with acid suppression was as efficient as a standard dose of a PEP alone in patients with CF [58]. Also, the late activation of PEPs in the more distal parts of the bowel might be a contributing factor to stenoses and the thickening of the ascending colon in CF patients treated with high doses of PEPs [44, 59]. That uncoated PEPs have been equivalent in their digestive effectiveness compared to enteric-coated PEPs has been a consistent finding [34, 60], and concentrations of trypsin and lipase aspirated from the duodenojejunal junction have been found to be higher using uncoated PEPs [34]. A further elevation of duodenal enzyme delivery and reduction of steatorrhea was shown when acid-suppressing drugs were combined with uncoated PEPs [22]. The combined use of uncoated enzymes with bicarbonate buffering showed no improvement of symptoms in patients with CF [32, 35]. A combination of PPIs and coated and uncoated enzymes as a rational approach in patients with severe maldigestion and insufficient therapeutic success from conventional approaches, but currently no data on this topic exist.

Conclusion

A beneficial effect of the use of acid suppression in EPI due to CP and CF has been well documented, while adverse effects have been disproved. The addition of PPIs to oral substitution of PEPs in patients with EPI results in a higher duodenal pH and

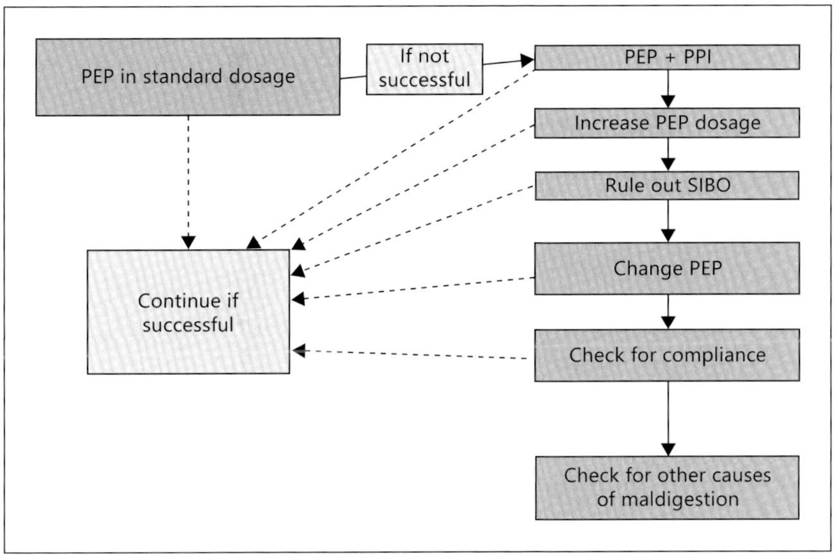

Fig. 2. Algorithm for treatment of EPI. SIBO = Small intestine bacterial overgrowth. PEP = Pancreatic enzyme preparation.

thereby a reduced inactivation of the residual enzymes secreted by the pancreas, an earlier dissolution of PEPs from the enteric coating, and a prevention of bile salt precipitation. Altogether, these factors contribute to a more effective and physiological digestion in patients with EPI. For this reason we suggest that PPIs have a crucial role in the therapeutic algorithm for EPI (fig. 2) and a generous recommendation for the use of PPIs in patients with EPI appears justified.

References

1 You CH, Rominger JM, Chey WY: Potentiation effect of cholecystokinin-octapeptide on pancreatic bicarbonate secretion stimulated by a physiologic dose of secretin in humans. Gastroenterology 1983;85:40–45.
2 Verkijk M, Gielkens HA, Lamers CB, Masclee AA: Effect of gastrin on antroduodenal motility: role of intraluminal acidity. Am J Physiol 1998;275:G1209–G1216.
3 Holtmann G, Kelly DG, Sternby B, DiMagno EP: Survival of human pancreatic enzymes during small bowel transit: effect of nutrients, bile acids, and enzymes. Am J Physiol 1997;273:G553–G558.
4 Keim V, Witt H, Bauer N, Bodeker H, Rosendahl J, Teich N, et al: The course of genetically determined chronic pancreatitis. JOP 2003;4:146–154.
5 Pelley JR, Gordon SR, Gardner TB: Abnormal duodenal [HCO3-] following secretin stimulation develops sooner than endocrine insufficiency in minimal change chronic pancreatitis. Pancreas 2012;41:481–484.
6 Ammann RW, Muellhaupt B, Meyenberger C, Heitz PU: Alcoholic nonprogressive chronic pancreatitis: prospective long-term study of a large cohort with alcoholic acute pancreatitis (1976–1992). Pancreas 1994;9:365–373.
7 Layer PH, DiMagno EP: Natural histories of alcoholic and idiopathic chronic pancreatitis. Pancreas 1996;12:318–320.
8 Robinson PJ, Smith AL, Sly PD: Duodenal pH in cystic fibrosis and its relationship to fat malabsorption. Dig Dis Sci 1990;35:1299–1304.
9 Youngberg CA, Berardi RR, Howatt WF, Hyneck ML, Amidon GL, Meyer JH, et al: Comparison of gastrointestinal pH in cystic fibrosis and healthy subjects. Dig Dis Sci 1987;32:472–480.
10 Zentler-Munro PL, Fine DR, Batten JC, Northfield TC: Effect of cimetidine on enzyme inactivation, bile acid precipitation, and lipid solubilisation in pancreatic steatorrhoea due to cystic fibrosis. Gut 1985;26:892–901.

11 Tran TM, Van den NA, Hendriks JJ, Forget P, Forget PP: Effects of a proton-pump inhibitor in cystic fibrosis. Acta Paediatr 1998;87:553–558.
12 DiMagno EP, Go VL, Summerskill WH: Relations between pancreatic enzyme ouputs and malabsorption in severe pancreatic insufficiency. N Engl J Med 1973;288:813–815.
13 Nakamura T, Takebe K, Imamura K, Tando Y, Yamada N, Arai Y, et al: Fat-soluble vitamins in patients with chronic pancreatitis (pancreatic insufficiency). Acta Gastroenterol Belg 1996;59:10–14.
14 Dutta SK, Bustin MP, Russell RM, Costa BS: Deficiency of fat-soluble vitamins in treated patients with pancreatic insufficiency. Ann Intern Med 1982;97: 549–552.
15 Layer P, Holtmann G: Pancreatic enzymes in chronic pancreatitis. Int J Pancreatol 1994;15:1–11.
16 Layer P, der Ohe MR, Holst JJ, Jansen JB, Grandt D, Holtmann G, et al: Altered postprandial motility in chronic pancreatitis: role of malabsorption. Gastroenterology 1997;112:1624–1634.
17 Keller J, Runzi M, Goebell H, Layer P: Duodenal and ileal nutrient deliveries regulate human intestinal motor and pancreatic responses to a meal. Am J Physiol 1997;272:G632–G637.
18 Mancilla AC, Madrid SAM, Hurtado HC, Orellana BC, Pena ZM, Tobar AE, et al: Small intestine bacterial overgrowth in patients with chronic pancreatitis (in Spanish). Rev Med Chil 2008;136:976–980.
19 Caspary WF, Winckler K, Lankisch PG, Creutzfeldt W: Influence of exocrine and endocrine pancreatic function on intestinal brush border enaymatic activities. Gut 1975;16:89–92.
20 Malfertheiner P, Dominguez-Munoz JE: Effect of exogenous pancreatic enzymes on gastrointestinal and pancreatic hormone release and gastrointestinal motility. Digestion 1993;54(suppl 2):15–20.
21 Glasbrenner B, Malfertheiner P, Pieramico O, Klatt S, Riepl R, Friess H, et al: Gallbladder dynamics in chronic pancreatitis. Relationship to exocrine pancreatic function, CCK, and PP release. Dig Dis Sci 1993;38:482–489.
22 Regan PT, Malagelada JR, DiMagno EP, Glanzman SL, Go VL: Comparative effects of antacids, cimetidine and enteric coating on the therapeutic response to oral enzymes in severe pancreatic insufficiency. N Engl J Med 1977;297:854–858.
23 Bovo P, Cataudella G, Di Francesco V, Vaona B, Filippini M, Marcori M, et al: Intraluminal gastric pH in chronic pancreatitis. Gut 1995;36:294–298.
24 Dutta SK, Russell RM, Iber FL: Impaired acid neutralization in the duodenum in pancreatic insufficiency. Dig Dis Sci 1979;24:775–780.
25 Regan PT, Malagelada JR, DiMagno EP, Go VL: Postprandial gastric function in pancreatic insufficiency. Gut 1979;20:249–254.
26 Saunders JH, Cargill JM, Wormsley KG: Gastric secretion of acid in patients with pancreatic disease. Digestion 1978;17:365–369.
27 Geus WP, Eddes EH, Gielkens HA, Gan KH, Lamers CB, Masclee AA: Post-prandial intragastric and duodenal acidity are increased in patients with chronic pancreatitis. Aliment Pharmacol Ther 1999;13:937–943.
28 Dimagno EP, Malagelada JR, Go VL, Moertel CG: Fate of orally ingested enzymes in pancreatic insufficiency. Comparison of two dosage schedules. N Engl J Med 1977;296:1318–1322.
29 Guarner L, Rodriguez R, Guarner F, Malagelada JR: Fate of oral enzymes in pancreatic insufficiency. Gut 1993;34:708–712.
30 Layer P, Groger G: Fate of pancreatic enzymes in the human intestinal lumen in health and pancreatic insufficiency. Digestion 1993;54(suppl 2):10–14.
31 Zentler-Munro PL, Fitzpatrick WJ, Batten JC, Northfield TC: Effect of intrajejunal acidity on aqueous phase bile acid and lipid concentrations in pancreatic steatorrhoea due to cystic fibrosis. Gut 1984;25:500–507.
32 Kalnins D, Corey M, Ellis L, Durie PR, Pencharz PB: Combining unprotected pancreatic enzymes with pH-sensitive enteric-coated microspheres does not improve nutrient digestion in patients with cystic fibrosis. J Pediatr 2005;146:489–493.
33 Gan KH, Heijerman HG, Geus WP, Bakker W, Lamers CB: Comparison of a high lipase pancreatic enzyme extract with a regular pancreatin preparation in adult cystic fibrosis patients. Aliment Pharmacol Ther 1994;8:603–607.
34 Dutta SK, Rubin J, Harvey J: Comparative evaluation of the therapeutic efficacy of a pH-sensitive enteric coated pancreatic enzyme preparation with conventional pancreatic enzyme therapy in the treatment of exocrine pancreatic insufficiency. Gastroenterology 1983;84:476–482.
35 Kalnins D, Ellis L, Corey M, Pencharz PB, Stewart C, Tullis E, et al: Enteric-coated pancreatic enzyme with bicarbonate is equal to standard enteric-coated enzyme in treating malabsorption in cystic fibrosis. J Pediatr Gastroenterol Nutr 2006;42:256–261.
36 Gan KH, Geus WP, Bakker W, Lamers CB, Heijerman HG: In vitro dissolution profiles of enteric-coated microsphere/microtablet pancreatin preparations at different pH values. Aliment Pharmacol Ther 1996;10:771–775.
37 Lohr JM, Hummel FM, Pirilis KT, Steinkamp G, Korner A, Henniges F: Properties of different pancreatin preparations used in pancreatic exocrine insufficiency. Eur J Gastroenterol Hepatol 2009;21:1024–1031.
38 Halm U, Loser C, Lohr M, Katschinski M, Mossner J: A double-blind, randomized, multicentre, crossover study to prove equivalence of pancreatin minimicrospheres versus microspheres in exocrine pancreatic insufficiency. Aliment Pharmacol Ther 1999; 13:951–957.

39 Giuliano CA, Dehoorne-Smith ML, Kale-Pradhan PB: Pancreatic enzyme products: digesting the changes. Ann Pharmacother 2011;45:658–666.
40 Stallings VA, Stark LJ, Robinson KA, Feranchak AP, Quinton H: Evidence-based practice recommendations for nutrition-related management of children and adults with cystic fibrosis and pancreatic insufficiency: results of a systematic review. J Am Diet Assoc 2008;108:832–839.
41 Layer P, Keller J: Lipase supplementation therapy: standards, alternatives, and perspectives. Pancreas 2003;26:1–7.
42 Sikkens EC, Cahen DL, van Eijck C, Kuipers EJ, Bruno MJ: Patients with exocrine insufficiency due to chronic pancreatitis are undertreated: a Dutch national survey. Pancreatology 2012;12:71–73.
43 Smyth RL, van Velzen D, Smyth AR, Lloyd DA, Heaf DP: Strictures of ascending colon in cystic fibrosis and high-strength pancreatic enzymes. Lancet 1994; 343:85–86.
44 Lebenthal E: High strength pancreatic exocrine enzyme capsules associated with colonic strictures in patients with cystic fibrosis: 'more is not necessarily better'. J Pediatr Gastroenterol Nutr 1994;18:423–425.
45 Brown A, Hughes M, Tenner S, Banks PA: Does pancreatic enzyme supplementation reduce pain in patients with chronic pancreatitis: a meta-analysis. Am J Gastroenterol 1997;92:2032–2035.
46 Garces MC, Gomez-Cerezo J, Codoceo R, Grande C, Barbado J, Vazquez JJ: Postprandial cholecystokinin response in patients with chronic pancreatitis in treatment with oral substitutive pancreatic enzymes. Dig Dis Sci 1998;43:562–566.
47 Mossner J, Keim V: Pancreatic enzyme therapy. Dtsch Arztebl Int 2010;108:578–582.
48 Ratuapli SK, Ellington TG, O'Neill MT, Umar SB, Harris LA, Foxx-Orenstein AE, et al: Proton pump inhibitor therapy use does not predispose to small intestinal bacterial overgrowth. Am J Gastroenterol 2012;107:730–735.
49 Choung RS, Ruff KC, Malhotra A, Herrick L, Locke GR III, Harmsen WS, et al: Clinical predictors of small intestinal bacterial overgrowth by duodenal aspirate culture. Aliment Pharmacol Ther 2011;33: 1059–1067.
50 Gan KH, Geus WP, Lamers CB, Heijerman HG: Effect of omeprazole 40 mg once daily on intraduodenal and intragastric pH in *H. pylori*-negative healthy subjects. Dig Dis Sci 1997;42:2304–2309.
51 Cai J, Zhou W, Luo HS, Peng LV: Effect of proton pump inhibitor on amylase release from isolated pancreatic acini. In Vitro Cell Dev Biol Anim 2007; 43:25–27.
52 Bruno MJ, Rauws EA, Hoek FJ, Tytgat GN: Comparative effects of adjuvant cimetidine and omeprazole during pancreatic enzyme replacement therapy. Dig Dis Sci 1994;39:988–992.
53 DiMagno EP: Gastric acid suppression and treatment of severe exocrine pancreatic insufficiency. Best Pract Res Clin Gastroenterol 2001;15:477–486.
54 Nakamura T, Arai Y, Tando Y, Terada A, Yamada N, Tsujino M, et al: Effect of omeprazole on changes in gastric and upper small intestine pH levels in patients with chronic pancreatitis. Clin Ther 1995;17: 448–459.
55 Francisco MP, Wagner MH, Sherman JM, Theriaque D, Bowser E, Novak DA: Ranitidine and omeprazole as adjuvant therapy to pancrelipase to improve fat absorption in patients with cystic fibrosis. J Pediatr Gastroenterol Nutr 2002;35:79–83.
56 Heijerman HG, Lamers CB, Bakker W, Dijkman JH: Improvement of fecal fat excretion after addition of omeprazole to pancrease in cystic fibrosis is related to residual exocrine function of the pancreas. Dig Dis Sci 1993;38:1–6.
57 Heijerman HG, Lamers CB, Bakker W: Omeprazole enhances the efficacy of pancreatin (pancrease) in cystic fibrosis. Ann Intern Med 1991;114:200–201.
58 Vecht J, Symersky T, Lamers CB, Masclee AA: Efficacy of lower than standard doses of pancreatic enzyme supplementation therapy during acid inhibition in patients with pancreatic exocrine insufficiency. J Clin Gastroenterol 2006;40:721–725.
59 Smyth RL, van Velzen D, Smyth AR, Lloyd DA, Heaf DP: Strictures of ascending colon in cystic fibrosis and high-strength pancreatic enzymes. Lancet 1994; 343:85–86.
60 Gouerou H, Dain MP, Parrondo I, Poisson D, Bernades P: Alipase versus nonenteric-coated enzymes in pancreatic insufficiency. A French multicenter crossover comparative study. Int J Pancreatol 1989; 5(suppl):45–50.

Prof. P. Malfertheiner
Department of Gastroenterology, Hepatology and Infectious Diseases
Otto von Guericke University, Leipziger Strasse 44
DE–39120 Magdeburg (Germany)
E-Mail peter.malfertheimer@med.ovgu.de

Chiba T, Malfertheiner P, Satoh H (eds): Proton Pump Inhibitors: A Balanced View.
Front Gastrointest Res. Basel, Karger, 2013, vol 32, pp 92–101 (DOI: 10.1159/000350635)

Proton Pump Inhibitors and Gastrointestinal Side Effects

Shotaro Nakamura[a] · Hiroshi Seno[b] · Tsutomu Chiba[b]

[a]Center for Advanced Medical Innovation, Kyushu University, Fukuoka, and [b]Department of Gastroenterology and Hepatology, Graduate School of Medicine, Kyoto University, Kyoto, Japan

Abstract

Proton-pump inhibitors (PPIs) are some of the most widely used drugs in the world. Although the safety of PPIs is well established, there are several side effects, including some which have only recently been recognized. The side effects include diarrhea and various symptoms due to overgrowth of pathogenic and nonpathogenic bacteria, rebound acid hypersecretion (RAHS), microscopic colitis, and the development of fundic gland polyps. The possibility of developing enterochromaffin-like carcinoid tumor and gastric cancer by PPI administration has been extensively discussed, but needs further investigation. Although the association of long-term PPI use with aggravation of atrophic gastritis is still controversial, *Helicobacter pylori* eradication may be recommended for patients receiving long-term PPI therapy. Finally, since PPI use is now extremely common, the dangerous consequences of RAHS, such as intestinal perforation or dehydration, raises caution that PPI therapy should not be stopped abruptly in patients with acid hypersecretion.

Rebound Acid Hypersecretion

Rebound acid hypersecretion (RAHS) is defined as an increase in gastric acid secretion above pretreatment levels following antisecretory therapy [1]. Proton pump inhibitor (PPI)-induced sustained hypergastrinemia provokes hyperplasia of histamine-secreting enterochromaffin-like (ECL) cells and subsequently increases parietal cell mass. This increase in cell mass precipitates an increase in the gastric acid secretory capacity that is not obvious during PPI therapy. Discontinuation of PPI therapy in such a condition may often cause rebound hypersecretion of gastric acid.

Indeed, previous studies on acid secretion following 8 weeks to 3 months of PPI therapy have shown that compared with the baseline levels, an increase in both basal and maximal acid levels was observed 14 days after treatment [2–4]. Waldum et al. [2] compared gastric acid secretion before omeprazole administration and 14 days after withdrawal of the drug following a 3-month therapy in patients with gastroesophageal reflux disease. The findings revealed a significant increase in both basal and pentagastrin-stimulated gastric acid secretion following omeprazole therapy. Gillen et al. [3, 4] examined acid secretion before, during, and 15 days after an 8-week therapy of omeprazole in healthy volunteers. Compared with pretreatment levels, they observed an 82 and 28% increase in basal and maximal acid secretion, respectively, 15 days after discontinuing the therapy in *Helicobacter pylori*-negative subjects. More recently, Reimer et al. [5] evaluated whether PPI discontinuation has any clinical implications in a double-blind placebo-controlled study of 120 healthy volunteers. Subjects were randomized to 8 weeks of placebo or 40 mg/day of esomeprazole followed by 4 weeks of placebo. The findings revealed that PPI therapy for only 8 weeks induced acid-related symptoms in healthy subjects after withdrawal; these symptoms may continue for at least 4 weeks after discontinuing the therapy. Another randomized controlled study conducted by Niklasson et al. [6] revealed that a 4-week pantoprazole therapy (40 mg/day) in *H. pylori*-negative healthy subjects significantly increased dyspeptic symptoms for 2 weeks after discontinuation compared with placebo.

Because *H. pylori* infection is associated with decreased capacity of acid secretion in patients with atrophic gastritis, *H. pylori* status might affect the rebound phenomenon. Previous studies have clearly shown that RAHS occurs after PPI discontinuation in *H. pylori*-negative subjects [2, 3, 6]. However, whether RAHS develops in *H. pylori*-positive subjects remains controversial. At present, it appears reasonable to conclude that RAHS has clinical implications, particularly in patients without *H. pylori* infection.

Attention should be paid to the dangerous consequences of discontinuing PPI therapy, particularly in patients with Zollinger-Ellison syndrome (ZES) who have been stabilized with a high PPI dose. Recently, Poitras et al. [7] reported 2 patients with ZES who developed severe complications immediately after PPI discontinuation. One patient developed multiple strictures of the esophagus, which required repeated dilatations over several years, and the other developed microperforations due to duodenal ulcers. Notably, both patients discontinued PPI therapy to proceed to further diagnostic testing; they developed complications within 48 h after discontinuing PPI therapy. In addition, we encountered 2 patients with ZES who developed severe dehydration and renal dysfunction due to vigorous diarrhea and vomiting after abrupt discontinuation of PPIs (unpubl. data). These findings indicate that PPIs should be continued in patients with proven or suspected ZES throughout the diagnostic evaluation period.

An increase in the number of patients with RAHS is anticipated because PPI use is extremely common, and it continues to escalate not only for patients with acid-related diseases, but also for those with undiagnosed symptoms. Therefore, novel methods for managing RAHS are eagerly awaited. One method could be tapering the dose of PPIs rather than abruptly discontinuing PPIs; however, the efficacy of this method has not yet been proven. Because PPIs induce hypergastrinemia with a resulting increase in histamine release, using histamine H_2-receptor antagonists (H_2-RAs) could be an alternative method, although development of tolerance to H_2-RAs during PPI therapy has been suggested. Finally, recognition of the prevalence of RAHS raises caution that PPIs should not be readily prescribed in patients with undiagnosed symptoms.

Diarrhea and Enteric Bacteria

Diarrhea is a common adverse effect of long-term PPI therapy [8–10]. Although specific causes of diarrhea have not been determined, it may develop because of enteric infections. Gastric acid functions as a chemical barrier that prevents orally ingested bacteria from reaching the intestine. Gastric pH <4.0 has a powerful bactericidal effect on exogenous acid-sensitive bacteria [11]. Once gastric pH is elevated, ingested microbes may disrupt gut homeostasis [12, 13]. Recent studies have provided epidemiological data to support this possibility.

Pathogenic bacteria have varying susceptibilities to gastric acid. *Campylobacter* and *Salmonella* are acid-sensitive microbes. Several studies have revealed an association between enteric infection caused by these two organisms and PPI use. A meta-analysis of 1,924 patients with enteric infections revealed that *Campylobacter* and *Salmonella* numbers were increased in users of acid inhibitors including PPI [odds ratio (OR): 2.55; 95% CI: 1.53–4.26], and that the association was greater in PPI use (OR: 3.33, 95% CI: 1.84–6.02) compared with H_2-RA use (OR: 2.03, 95% CI: 1.05–3.92) [14]. A recent systematic review also showed evidence for enhanced susceptibility to enteric infection caused by *Campylobacter* and *Salmonella* in PPI users, with adjusted relative risk ranges of 3.5–11.7 and 4.2–8.3, respectively [15]. Other pathogenic bacteria, such as *Escherichia coli*, *Shigella*, *Listeria*, and *Vibrio cholerae* are generally more acid-resistant than *Campylobacter* and *Salmonella*. However, several reports have suggested enhanced susceptibility to these organisms in patients using PPIs. Because of a lack of large-scale studies, further investigations are needed to specifically determine the association between PPI use and the susceptibility to these acid-resistant microbes.

Clostridium difficile (*C. difficile*)-associated diarrhea is a growing public health concern in hospitals and communities with or without traditional risk factors, such as the use of antibiotics. In addition, a growing body of evidence has shown that *C. difficile* infection is associated with PPI use. In 2012, the FDA announced that PPI use may be linked to an increased risk of *C. difficile*-associated diarrhea [16]. Spores of *C. difficile*, which are acid resistant, can be the primary vehicle for spread throughout environmental surfaces. In contrast, the vegetative form of *C. difficile*, which is acid sensitive, survives in the stomach only in high gastric pH. However, where and when the conversion of spores to the vegetative form occurs remains unclear.

Following several studies showing an association between PPI use and increased susceptibility to *C. difficile*-associated diarrhea, a meta-analysis involving 2,948 patients demonstrated that various durations of PPI therapy in hospitalized patients caused an increased risk of *C. difficile*-associated diarrhea (OR: 3.33, 95% CI: 1.84–6.02) [14]. A more recent large-scale meta-analysis including 202,965 patients indicated that PPI use increased the risk of *C. difficile* infection (OR: 2.15, 95% CI: 1.81–2.55), although a significant heterogeneity among study results was observed [17]. Therefore, multiple studies have identified an association between PPI use and the development of community-acquired and healthcare-associated *C. difficile* infections. However, prospective studies confirming this association and careful consideration of the risks and benefits of PPI use, particularly in hospitalized patients, is required.

Small intestinal bacterial overgrowth (SIBO) is a clinical condition characterized by various symptoms that occur when the proximal small intestine is colonized by a large number (>10^5 colony-forming units/ml) of endogenous bacterial flora. Generally, gut microflora have metabolic, trophic, and protective effects on the host, thus regulating the immune system and the absorption of nutrients. However, elevated gastric pH affects the composition of not only the upper gut flora, but also of the lower intestinal microflora. Therefore, the association between SIBO and PPI use has recently generated clinical interest.

The data regarding the relevance of related clinical outcomes in patients with SIBO who have been treated with PPIs are controversial. A recent analysis involving 450 patients who were evaluated using the glucose hydrogen breath test showed that SIBO occurred more frequently in long-term PPI users than in controls and patients with irritable bowel syndrome (IBS) [18]. Another retrospective study involving 1,191 patients revealed that PPI use did not predispose to SIBO as determined by the glucose hydrogen breath test [19]. More than 10 studies, including the above two, have attempted to address the association between SIBO and PPI use, and have reported conflicting conclusions. The lack of a 'gold standard' test to diagnose SIBO [20] could be the reason for this discrepancy. In addition, it is unclear whether SIBO is involved in IBS associated with PPI use. Of note, gastroesophageal reflux disease, which is the most frequent disorder treated with PPI, is associated with a high incidence of IBS, and vice versa. Although studies indicate that SIBO is prevalent in IBS patients, a causal link between SIBO and IBS has not been established [21, 22]. Further prospective studies are required to confirm the association between SIBO, IBS, and PPI use.

The association between nonsteroidal anti-inflammatory drugs (NSAIDs) and gastric and duodenal mucosal damage is well known. In addition, there is a report showing that approximately 70% of chronic NSAID users exhibited mucosal damage in the small intestine [23]. PPIs decrease the incidence of NSAID-induced gastroduodenal damage [24]; therefore, PPIs and NSAIDs are frequently prescribed simultaneously to protect against gastroduodenal injury caused by the NSAIDs. However, despite several recent studies that used video capsule endoscopy, there have been no large-scale studies demonstrating the protective effect of PPIs against NSAID-induced damage in the small intestine in humans. In contrast, a recent report on rodent models suggested that PPIs exacerbate NSAID-induced intestinal damage partly because of significant shifts in enteric microbial populations [25]. In addition, previous studies have consistently shown that enteric bacteria, particularly Gram-negative bacteria, significantly contribute to NSAID-induced mucosal damage in rodent intestine [26]. If PPI use is associated with SIBO, it may be reasonable to suspect that concomitant PPI and NSAID use results in a negative chain reaction in the protection of small intestinal mucosa. This association needs to be investigated in further prospective studies despite the technical difficulties associated with accurately evaluating dysbiosis and mucosal lesions in the small intestine.

Microscopic Colitis (Collagenous Colitis and Lymphocytic Colitis)

Microscopic colitis (MC) is a distinct clinicopathological entity characterized by chronic watery diarrhea and endoscopically normal or minimally abnormal colonic mucosa [27, 28]. Histologically, MC is divided into two subtypes: collagenous colitis and lymphocytic colitis. Collagenous colitis is characterized by a thickened subepithelial collagen band (>10 μm), whereas lymphocytic colitis is characterized by increased infiltration of intraepithelial lymphocytes (>20 lymphocytes/100 epithelial cells) [27, 28]. The clinical presentation and management strategy for both subtypes are the same. Both collagenous and lymphocytic colitis predominantly affect elderly women. Nonetheless, it is unclear whether collagenous colitis and lymphocytic colitis have the same pathophysiology.

In the last decade, epidemiological studies have shown a rising incidence of MC, which is as common as classical inflammatory bowel diseases (i.e. Crohn's disease and ulcerative colitis) [28]. The reported incidence rates after 2002 have been 2.6–10.8 cases/100,000 inhabitants for collagenous colitis and 2.2–14.0 cases/100,000 inhabitants for lymphocytic colitis [27, 28].

Although the pathogenesis of MC has not been elucidated, various drugs are considered to be responsible for the development of the disease [27–29]. One literature review described that NSAIDs, aspirin, ranitidine, selective serotonin reuptake inhibitors, ticlopidine, acarbose, statins, and PPIs (particularly lansoprazole) have high or intermediate levels of association with MC [29]. A case-control study revealed that exposure to PPIs was significantly more frequent in patients with MC than in controls [30]. Recently, the number of reported lansoprazole-related MC cases has been increasing [31, 32]. A review of 33 patients with lansoprazole-related MC revealed that endoscopic alterations (i.e. linear mucosal defects or erosions) were observed more frequently in collagenous colitis (13 of 18 cases, 72%) than in lymphocytic colitis (1 of 15 cases, 7%), and that the time period between the start of lansoprazole therapy and the onset of symptoms was longer in patients with collagenous colitis than in those with lymphocytic colitis (median: 60 vs. 28 days) [31]. These observations suggest differences in the inflammatory process involved in lansoprazole-related collagenous colitis and lymphocytic colitis.

The mechanisms by which lansoprazole induces MC remain uncertain. Proton pumps (HK-ATPases) are present in colonic epithelial cells [33]. Because colonic HK-ATPases are highly homologous to parietal cell HK-ATPases, lansoprazole may have a direct effect on the colonic epithelial cells, which may trigger an immune response causing MC [31, 32]. However, cases of MC associated with PPIs other than lansoprazole are rare; few cases of MC associated with omeprazole or esomeprazole have been reported to date [34]. The rarity of the association between MC and PPIs other than lansoprazole may be because of the difference in the affinity of lansoprazole and other PPIs to colonic HK-ATPases.

The diagnosis of MC is primarily based on the clinical symptoms and the characteristic histopathological findings of biopsy specimens from the colonic mucosa. Discontinuation of possible causative drugs and selecting the appropriate treatment on the basis of the severity of symptoms should be considered while managing patients with MC [27, 28]. In mild cases, nonspecific antidiarrheal drugs such as loperamide or cholestyramine followed by mesalazine or bismuth can be administered [27]. In moderate-to-severe diarrhea that does not respond to the aforementioned medications, oral administration of budesonide is recommended [27, 28]. Furthermore, in patients who do not respond to the above-mentioned drugs, prednisolone, immunomodulators such as azathioprine, 6-mercaptopurine, or anti-TNF-α antibodies can be considered.

Fundic Gland Polyps

Fundic gland polyps (FGPs) are the most common benign gastric polyps, histologically characterized by cystic dilatations of the oxyntic glands lined by variably flattened parietal and chief cells with or without mucous foveolar cells. These polyps are endoscopically detected in ≤1.9% of the general population and in ≤84% patients with familial adenomatous polyposis (FAP). Sporadic FGPs are predominantly detected in women and *H. pylori*-negative subjects. Epithelial dysplasia is observed in ≤25% FAP-associated FGPs, but is extremely rare in sporadic FGPs [35].

Long-term PPI use is reportedly associated with a greater than or equal to fourfold increase in the incidence of FGP [36–38]. A prospective multicenter study revealed that 14% of the patients developed new FGPs following rabeprazole therapy (10 mg/day) for 104 weeks, which was associated with a *H. pylori*-negative status, but not with hypergastrinemia [37]. Polyp regression has been observed after cessation of PPI therapy [39] or acquisition of *H. pylori* [40]. The possible mechanisms involved in the development of FGPs due to long-term PPI use are not well understood. However, PPIs inhibit the active release of hydrochloric acid from secretory canaliculi, which may not only increase the parietal cell mass, but also the parietal cell protrusions. Such protrusions may obstruct the isthmus, resulting in fundic gland cyst formation, which in turn may enlarge and develop into FGPs [38].

Patients with FAP syndrome develop both gastric adenomas and FGPs. Gastric adenoma is frequently detected in *H. pylori*-positive patients, whereas FGP has been predominantly detected in *H. pylori*-negative patients [41]. In the general population, adenoma is considered to be a precancerous lesion, whereas an FGP is considered to be a benign polyp. However, several reports have shown that an FGP in patients with FAP syndrome develops into gastric adenocarcinoma [42–44]. In this regard, adenomatous changes [45] or high-grade dysplasia [46] have also been observed in sporadic FGPs. In addition, a new autosomal dominant syndrome, gastric adenocarcinoma and proximal polyposis of the stomach or familial fundic gland polyposis, has been described recently [47, 48]. Notably, these families characteristically have numerous FGPs in the stomach and appear to be at risk for developing cancer from FGP. Some sporadic FGPs possess a mutation in the β-catenin gene [40, 49]; however, it is unclear whether FGPs induced by long-term PPI therapy possess this mutation. No instances of dysplasia in PPI-induced FGPs have been reported. Therefore, it is unclear whether occasional endoscopic examination is recommended in patients with many or large FGPs. It is also unknown whether PPI therapy in patients with FAP and gastric adenocarcinoma and proximal polyposis of the stomach/familial fundic gland polyposis increases their risk for progression of FGPs and dysplasia.

Development of Enterochromaffin-Like Cell Hyperplasia and Enterochromaffin-Like Carcinoid Tumor: Neuroendocrine Tumor G1

Gastrin is a trophic factor for ECL cells [50, 51], and patients with prolonged hypergastrinemia due to ZES or type A atrophic gastritis are predisposed to developing ECL cell hyperplasia and ECL carcinoid tumor (NET). Therefore, it is theoretically possible that prolonged hypergastrinemia due to long-term PPI therapy induces ECL cell hyperplasia or even ECL NET. In animal experiments, long-term administration of PPIs caused NET in rats, but not in mice [52]. In humans, long-term PPI therapy has been shown to be associated with an increase in argyrophil cell density and micronodular hyperplasia [53]. However, only a few case reports on the development of ECL NET or gastric neuroendocrine carcinoma during administration of PPIs have been noted; none of the cases reported thus far have clearly indicated that PPI therapy caused the tumor development [54–56]. This issue is discussed in the chapter by Satoh [pp. 1–17].

Progression of Chronic Atrophic Gastritis and Gastric Cancer

Since the use of PPIs in medicine before 1990, several reports have suggested that PPI use aggravates corpus gastritis in patients with *H. pylori* infection. Solcia et al. [57] reported increased gastritis activity in the corpus mucosa after short-term (4 weeks) use of PPIs in 65 patients with duodenal ulcers. Kuipers et al. [58] conducted a long-term observation with a mean follow-up period of 5 years, and suggested that after PPI treatment, *H. pylori*-positive patients with reflux esophagitis are at an increased risk of atrophic gastritis of the corpus compared with *H. pylori*-negative patients. This change in gastric corpus appears to be common among acid suppressants because long-term H_2-RA therapy has also been shown to aggravate corpus gastritis to a lesser extent than PPI use [59].

These reports have attracted much attention because the gastric mucosa of patients with intestinal-type gastric cancer is characterized by atrophic gastritis associated with intestinal metaplasia [60]. In Mongolian gerbils infected with *H. pylori*, 6-month PPI therapy aggravated atrophic corpus gastritis and promoted adenocarcinoma development [61]. However, in humans, the possibility that long-term PPI therapy is significantly associated with the development of atrophic gastritis remains controversial. No increase in intestinal metaplasia was observed on endoscopy performed during a 3-year PPI therapy [62]. Moreover, no existing cohort or large-scale studies have demonstrated an increased risk of gastric cancer in *H. pylori*-positive patients treated with acid suppressants, including PPIs. A prospective cohort study that followed-up 230 patients for up to 11 years failed to show an increased risk of gastric cancer with PPI use [63]. These matters are described in the chapter by Satoh [pp. 1–17].

References

1 US Food and Drug Administration: Rebound of gastric acid secretion. 2000. http://www.fda.gov/ohrms/dockets/ac/00/backgrd/3650b1a_11.pdf (accessed March 17, 2009).

2 Waldum HL, Arnestad JS, Brenna E, et al: Marked increase in gastric acid secretory capacity after omeprazol treatment. Gut 1996;39:649–653.

3 Gillen D, Wirz AA, Ardill JE, et al: Rebound hypersecretion after omeprazole and its relation to on-treatment acid suppression and *Helicobacter pylori* status. Gastroenterology 1999;116:239–247.

4 Gillen D, Wirz AA, Neithercut WD, et al: *Helicobacter pylori* infection potentiates the inhibition of gastric acid secretion by omeprazole. Gut 1999;44:468–475.

5 Reimer C, Sondergaard B, Hilsted L, et al: Proton-pump inhibitor therapy induces acid-related symptoms in healthy volunteers after withdrawal of therapy. Gastroenterology 2009;137:80–87.

6 Niklasson A, Lindstrom L, Simren M, et al: Dyspeptic symptom development after discontinuation of a proton pump inhibitor: a double-blind placebo-controlled trial. Am J Gastroenterol 2010;105:1531–1537.

7 Poitras P, Gingras MH, Rehfeld JF: The Zollinger-Ellison syndrome: dangers and consequences of interrupting antisecretory treatment. Clin Gastroenterol Hepatol 2012;10:199–202.

8 Leufkens H, Claessens A, Heerdink E, et al: A prospective follow-up study of 5669 users of lansoprazole in daily practice. Aliment Pharmacol Ther 1997;11:887–897.

9 Claessens AA, Heerdink ER, van Eijk JT, et al: Characteristics of diarrhoea in 10,008 users of lansoprazole in daily practice: which co-factors contribute? Pharmacoepidemiol Drug Saf 2002;11:703–708.

10 Pilotto A, Franceschi M, Vitale D, et al; FIRI (Fondazione Italiana Ricerca sull'Invecchiamento), SOFIA Project Investigators: The prevalence of diarrhea and its association with drug use in elderly outpatients: a multicenter study. Am J Gastroenterol 2008; 103:2816–2823.
11 Giannella RA, Broitman SA, Zamcheck N: Gastric acid barrier to ingested microorganisms in man: studies in vivo and in vitro. Gut 1972;13:251–256.
12 Gray JD, Shinner M: Influence of gastric pH on gastric and jejunal flora. Gut 1967;8:574–581.
13 Howden CW, Hunt RH: Relationship between gastric secretion and infection. Gut 1987;28:96–107.
14 Leonard J, Marshall JK, Moayyedi P: Systematic review of the risk of enteric infection in patients taking acid suppression. Am J Gastroenterol 2007;102:2047–2056, quiz 2057.
15 Bavishi C, Dupont HL: Systematic review: the use of proton pump inhibitors and increased susceptibility to enteric infection. Aliment Pharmacol Ther 2011; 34:1269–1281.
16 US Food and Drug Administration: Rebound of gastric acid secretion. 2012. http://www.fda.gov/Drugs/DrugSafety/ucm290510.htm.
17 Deshpande A, Pant C, Pasupuleti V, et al: Association between proton pump inhibitor therapy and *Clostridium difficile* infection in a meta-analysis. Clin Gastroenterol Hepatol 2012;10:225–233.
18 Lombardo L, Foti M, Ruggia O, Chiecchio A: Increased incidence of small intestinal bacterial overgrowth during proton pump inhibitor therapy. Clin Gastroenterol Hepatol 2010;8:504–508.
19 Ratuapli SK, Ellington TG, O'Neill MT, et al: Proton pump inhibitor therapy use does not predispose to small intestinal bacterial overgrowth. Am J Gastroenterol 2012;107:730–735.
20 Dukowicz AC, Lacy BE, Levine GM: Small intestinal bacterial overgrowth: a comprehensive review. Gastroenterol Hepatol (NY) 2007;3:112–122.
21 Spiegel BM, Chey WD, Chang L: Bacterial overgrowth and irritable bowel syndrome: unifying hypothesis or a spurious consequence of proton pump inhibitors? Am J Gastroenterol 2008;103:2972–2976.
22 Spiegel BM: Questioning the bacterial overgrowth hypothesis of irritable bowel syndrome: an epidemiologic and evolutionary perspective. Clin Gastroenterol Hepatol 2011;9:461–469, quiz e59.
23 Bjarnason I, Hayllar J, MacPherson AJ, et al: Side effects of nonsteroidal anti-inflammatory drugs on the small and large intestine in humans. Gastroenterology 1993;104:1832–1847.
24 Yeomans ND, Tulassay Z, Juhász L, et al: A comparison of omeprazole with ranitidine for ulcers associated with nonsteroidal antiinflammatory drugs. Acid Suppression Trial: Ranitidine versus Omeprazole for NSAID-Associated Ulcer Treatment (ASTRONAUT) Study Group. New Engl J Med 1998;338:719–726.
25 Wallace JL, Syer S, Denou E, et al: Proton pump inhibitors exacerbate NSAID-induced small intestinal injury by inducing dysbiosis. Gastroenterology 2011; 141:1314–1322, 1322.e1–e5.
26 Watanabe T, Higuchi K, Kobata A, et al: Non-steroidal anti-inflammatory drug-induced small intestinal damage is Toll-like receptor 4 dependent. Gut 2008; 57:181–187.
27 Pardi DS, Kelly CP: Microscopic colitis. Gastroenterology 2011;140:1155–1165.
28 Münch A, Aust D, Bohr J, et al; European Microscopic Colitis Group (EMCG): Microscopic colitis: current status, present and future challenges: statements of the European Microscopic Colitis Group. J Crohns Colitis 2012;6:932–945.
29 Beaugerie L, Pardi DS: Review article: drug-induced microscopic colitis – proposal for a scoring system and review of the literature. Aliment Pharmacol Ther 2005;22:277–284.
30 Keszthelyi D, Jansen SV, Schouten GA, de Kort S, et al: Proton pump inhibitor use is associated with an increased risk for microscopic colitis: a case-control study. Aliment Pharmacol Ther 2010;32:1124–1128.
31 Umeno J, Matsumoto T, Nakamura S, et al: Linear mucosal defect may be characteristic of lansoprazole-associated collagenous colitis. Gastrointest Endosc 2008;67:1185–1191.
32 Capurso G, Marignani M, Attilia F, et al: Lansoprazole-induced microscopic colitis: an increasing problem? Results of a prospecive case-series and systematic review of the literature. Dig Liver Dis 2011;43: 380–385.
33 Kaunitz JD, Sachs G: Identification of a vanadate-sensitive potassium-dependent proton pump from rabbit colon. J Biol Chem 1986;261:14005–14010.
34 Wilcox GM, Mattia AR: Microscopic colitis associated with omeprazole and esomeprazole exposure. J Clin Gastroenterol 2009;43:551–553.
35 Abraham SC, Park SJ, Mugartegui L, et al: Sporadic fundic gland polyps with epithelial dysplasia: evidence for preferential targeting for mutations in the adenomatous polyposis coli gene. Am J Pathol 2002; 161:1735–1742.
36 Jalving M, Koornstra JJ, Wesseling J, et al: Increased risk of fundic gland polyps during long-term proton pump inhibitor therapy. Aliment Pharmacol Ther 2006;24:1341–1348.
37 Hongo M, Fujimoto K, Gastric Polyps Study Group: Incidence and risk factor of fundic gland polyp and hyperplastic polyp in long-term proton pump inhibitor therapy: a prospective study in Japan. J Gastroenterol 2010;45:618–624.
38 Cats A, Schenk BE, Bloemena E, et al: Parietal cell protrusions and fundic gland cysts during omeprazole maintenance treatment. Hum Pathol 2000;31: 684–690.

39 Stolte M: Fundic gland polyps: a rare, innocuous, and reversible disturbance. Gastroenterology 1993; 105:1590–1591.
40 Watanabe N, Seno H, Nakajima T, et al: Regression of fundic gland polyps following acquisition of *Helicobacter pylori*. Gut 2002;51:742–745.
41 Nakamura S, Matsumoto T, Kobori Y, et al: Impact of *Helicobacter pylori* infection and mucosal atrophy on gastric lesions in patients with familial adenomatous polyposis. Gut 2002;51:485–489.
42 Zwick A, Munir M, Ryan CK, et al: Gastric adenocarcinoma and dysplasia in fundic gland polyps of a patient with attenuated adenomatous polyposis coli. Gastroenterology 1997;113:659–663.
43 Hofgartner WT, Thorp M, Ramus MW, et al: Gastric adenocarcinoma associated with fundic gland polyps in a patient with attenuated familial adenomatous polyposis. Am J Gastroenterol 1999;94:2275–2281.
44 Garrean S, Hering J, Saied A, et al: Gastric adenocarcinoma arising from fundic gland polyps in a patient with familial adenomatous polyposis syndrome. Am Surg 2008;74:79–83.
45 Tazaki S, Nozu F, Yosikawa N, et al: Sporadic fundic gland polyp-related adenomas occurred in non-atrophic gastric mucosa without *Helicobacter pylori* infection. Dig Endosc 2011;23:182–186.
46 Stolte M, Vieth M, Ebert MP: High-grade dysplasia in sporadic fundic gland polyps: clinically relevant or not? Eur J Gastroenterol Hepatol 2003;15:1153–1156.
47 Worthley DL, Phillips KD, Wayte N, et al: Gastric adenocarcinoma and proximal polyposis of the stomach (GAPPS): a new autosomal dominant syndrome. Gut 2012;61:774–779.
48 Yanaru-Fujisawa R, Nakamura S, Moriyama T, et al: Familial fundic gland polyposis with gastric cancer. Gut 2012;61:1103–1104.
49 Abraham SC, Nobukawa B, Giardiello FM, et al: Sporadic fundic gland polyps: common gastric polyps arising through activating mutations in the beta-catenin gene. Am J Pathol 2001;158:1005–1010.
50 Taniguchi T, Matsui T, Ito M, et al: Cholecystokinin-B/gastrin receptor signaling pathway involves tyrosine phosphorylations of p125FAK and p42MAP. Oncogene 1994;9:861–867.
51 Kinoshita Y, Nakata H, Kishi K, et al: Comparison of the signal transduction pathways activated by gastrin in enterochromaffin-like and parietal cells. Gastroenterology 1998;115:93–100.
52 Havu N: Enterochromaffin-like cell carcinoids of gastric mucosa in rats after life-long inhibition of gastric secretion. Digestion 1986;35(suppl 1):42–55.
53 Lamberts R, Creutzfeldt W, Struber HG, et al: Long-term omeprazole therapy in peptic ulcer disease: gastrin, endocrine cell growth, and gastritis. Gastroenterology 1993;104:1356–1370.
54 Dawson R, Manson JM: Omeprazole in oesophageal reflux disease. Lancet 2000;356:1770–1771.
55 Haga Y, Nakatsuura T, Shibata Y, et al: Human gastric carcinoid tumor during long term antiulcer therapy of H_2 receptor antagonist and proton pump inhibitor. Dig Dis Sci 1998;43:253–257.
56 Jlanu CS, Lange OJ, Viset T, et al: Gastric neuroendocrine carcinoma after long-term use of proton pump inhibitor. Scand J Gastroenterol 2012;47:64–67.
57 Solcia E, Villani L, Fiocca R, et al: Effects of eradication of *Helicobacter pylori* on gastritis in duodenal ulcer patients. Scand J Gastroenterol Suppl 1994;201:28–34.
58 Kuipers EJ, Lundell L, Klinkenberg-Knol EC, et al: Atrophic gastritis and *Helicobacter pylori* infection in patients with reflux esophagitis treated with omeprazole or fundoplication. New Engl J Med 1996;334: 1018–1022.
59 Meining A, Bosseckert H, Caspary WF, et al: H_2-receptor antagonists and antacids have an aggravating effect on *Helicobacter pylori* gastritis in duodenal ulcer patients. Aliment Pharmacol Ther 1997;11:729–734.
60 Correa P, Shiao YH: Phenotypic and genotypic events in gastric carcinogenesis. Cancer Res 1994;54(7 suppl):1941s–1943s.
61 Hagiwara T, Mukaisho K, Nakayama T, et al: Long-term proton pump inhibitor administration worsens atrophic corpus gastritis and promotes adenocarcinoma development in Mongolian gerbils infected with *Helicobacter pylori*. Gut 2011;60:624–630.
62 Uemura N, Okamoto S, Yamamoto S, et al: Changes in *Helicobacter pylori*-induced gastritis in the antrum and corpus during long-term acid-suppressive treatment in Japan. Aliment Pharmacol Ther 2000; 14:1345–1352.
63 Klinkenberg-Knol EC, Nelis F, Dent J, et al; Long-Term Study Group: Long-term omeprazole treatment in resistant gastroesophageal reflux disease: efficacy, safety, and influence on gastric mucosa. Gastroenterology 2000;118:661–669.

Prof. Tsutomu Chiba
Department of Gastroenterology and Hepatology
Graduate School of Medicine, Kyoto University
54 Kawahara-cho, Shogoin, Sakyo-ku, Kyoto 606-8507 (Japan)
E-Mail chiba@kuhp.kyoto-u.ac.jp

Chiba T, Malfertheiner P, Satoh H (eds): Proton Pump Inhibitors: A Balanced View.
Front Gastrointest Res. Basel, Karger, 2013, vol 32, pp 102–115 (DOI: 10.1159/000350638)

Administration of Proton Pump Inhibitors and Risk of Systemic Side Effects

Yoshikazu Kinoshita · Shunji Ishihara

Department of Gastroenterology and Hepatology, Shimane University School of Medicine, Izumo, Japan

Abstract

Proton pump inhibitors (PPIs) are widely used for maintenance treatment of gastroesophageal reflux diseases as well as prevention of gastroduodenal ulcers in patients taking nonsteroidal anti-inflammatory drugs. Since they are used for long periods concurrently with other drugs, potential side effects and drug interactions have received attention, with recently reported systemic side effects of PPIs found to be related to their potent acid inhibitory effect. Gastric acidity lowered by PPI administration may also be associated with decreased calcium absorption and cause osteoporosis, resulting in bone fracture. All previous reports concerning the risk of fracture in patients taking PPIs are retrospective observational studies and their results disagree, thus large prospective studies are necessary to clarify the risk of their administration for development of bone fracture. Vitamin B_{12} and iron absorption may also be decreased by acid inhibition. However, investigations of PPI-induced vitamin B_{12} and/or iron deficiency are scarce, and their results are inconsistent, though PPI-related risk does not seem to be high. Acid inhibition may also cause bacterial colonization in the distal esophagus and stomach. Reflux and aspiration of bacteria can lead to pneumonia in patients taking PPIs, and early retrospective studies suggested significant risks associated with their administration not only for community-acquired, but also hospital-acquired, pneumonia. However, prospective controlled studies have not confirmed those findings. Therefore, the risk for pneumonia associated with PPI use remains unresolved and may not be clinically meaningful. PPIs interact with other coadministered drugs at the steps of absorption and metabolism. Many drugs have their solubility and stability changed in solutions with different acidity. Therefore, acid inhibition easily alters the bioavailability of drugs such as digoxin, alendronate, and atazanavir. PPIs also influence the activation and degradation of coadministered drugs by competitively inhibiting the hepatic cytochrome p450 enzyme CYP2C19. For example, clopidogrel, an antithrombotic drug, requires activation by cytochrome p450 enzymes including CYP2C19, while PPIs inhibit the activation of clopidogrel and decrease its antithrombotic effect. However, the clinical significance of these actions has not been clearly shown. The systemic side effects of PPIs are equal to or less than those of other commonly used drugs. However, it is important to always note the appropriate indications for PPIs and carefully watch for possible systemic side effects even when the frequency is not high.

Table 1. Classification of possible side effects of PPIs

1. *Nonpredictable side effects*
Allergic drug reaction
Nonallergic drug reaction
hypomagnesemia, collagenous colitis
2. *Predictable side effects*
Related to gastric acid suppression
Decreased iron absorption
anemia
Decreased vitamin B_{12} absorption
anemia, neuropathy
Decreased calcium absorption
bone loss, fracture
Increased serum gastrin
colon cancer, carcinoid tumor
Increased *H. pylori*-related inflammation
gastric cancer
Decreased gastric bactericidal effect
pneumonia, enteric bacterial infection
3. *Drug interactions*
Related to gastric acid suppression
Increased bioavailability
digoxin, methyldigoxin
Decreased bioavailability
atazanavir, alendronate, etc.
Related to metabolism at cytochrome p450
Increased bioavailability
diazepam, phenytoin, etc.
Decreased bioavailability
clopidogrel

Proton pump inhibitors (PPIs) are widely used potent antisecretory drugs. Previously, they were mainly administered for short periods to heal peptic ulcers and esophageal erosions, as well as eradicate *Helicobacter pylori* infection. Now, however, they are often prescribed for longer periods concurrently with other drugs to maintain remission of gastroesophageal reflux disease and prevent nonsteroidal anti-inflammatory drug- or low-dose aspirin-induced gastroduodenal ulcers.

Since PPIs potently inhibit gastric acid secretion, they are also expected to influence gastric bactericidal effects, as well as digestion and absorption of various nutrients (table 1). In addition, inhibited gastric acid secretion is thought to change the intragastric degradation and solubility of ingested medicines, and alter their absorption efficiency. The majority of PPIs are degraded by CYP2C19, a hepatic drug metabolizing cytochrome p450 enzyme. Since many drugs are metabolized by this enzyme, PPIs may interact with the metabolism of other coadministered drugs at the enzyme level. Possible digestive and extradigestive (systemic) side effects that may occur during long-term administration, along with potential drug interactions, are emerging as important complications of treatment with PPIs. In this chapter, recently reported findings of extradigestive and systemic side effects related to the use of PPIs are summarized.

Impaired Absorption of Nutrients

Decreased Calcium Absorption and Bone Fracture
Osteoporosis and osteoporosis-related increases in bone fractures are common problems in elderly populations, with adequate intake and effective absorption of calcium known to be important to maintain normal bone integrity. Absorption of administered calcium carbonate has been reported to decrease under acid suppression by administration of PPIs [1] and their use is considered to decrease gastrointestinal calcium absorption from meals, in spite of conflicting findings [2]. When calcium absorption is decreased, inadequate mineralization of bone may occur and the risk of bone fracture can increase. In addition, PPI administration may decrease vitamin B_{12} absorption via decreased protein digestion in the stomach and also induce hyperparathyroidism through hypergastrinemia. Vitamin B_{12} deficiency and hyperparathyroidism can cause osteoblast and osteoclast dysfunction, with a concomitant decrease in the strength of bone structures. Therefore, the possible risk of bone fracture has been repeatedly investigated in patients treated with PPIs.

All studies of the risk of bone fracture during long-term treatment with PPIs are observational and the majority are retrospective in nature. Two case control studies of the risk of fracture during long-term PPI administration were reported in 2006, and their results showed that such administration increased the risk of fractures [3, 4]. Following those, several long-term observational studies presented both positive and negative risk factors for bone fractures related to PPI use (table 2) [3–15]. Recently, four different systematic reviews and meta-analyses of over ten partially different observational studies were presented [16–19], and showed that the risks of hip and spine fractures in PPI users are significantly higher than those in nonusers; however, dose-dependent risk was not given. Interestingly, PPI administration was not a statistically significant risk factor for hip fracture in long-term users, while it was found to be a significant risk factor in short-term users who had received PPI administration for less than 1 year. In addition, PPIs were not found to be a risk factor for wrist/forearm fractures. These apparent inconsistencies have raised concerns regarding the reliability of reported observational studies of fracture risk related to PPIs.

Targownik et al. [20, 21] reported changes of bone mineral density in longitudinal cohort studies. They found no significant decrease of bone mineral density in the PPI-treated group during 2- and 10-year observation periods, even when the dose given was greater than the standard. Since all studies of the risks of long-term PPI administration for fracture risks are observational in nature, the possibility of residual con-

Table 2. PPI administration and reported risks of fracture

Study	Study design	n	Fracture site	RR for PPI (95% CI)
Yang et al. [3], 2006	case control	148,942	hip	1.44 (1.30–1.59)
Vestergaard et al. [4], 2006	case control	498,617	hip	1.45 (1.28–1.65)
			spine	1.60 (1.25–2.04)
			any	1.18 (1.12–1.43)
Yu et al. [5], 2008	cohort	5,339	hip	1.16 (0.80–1.67)
			non-spine	1.34 (1.10–1.64)
Yu et al. [5], 2008	cohort	5,755	hip	0.62 (0.26–1.44)
			non-spine	1.21 (0.91–1.62)
Targownik et al. [6], 2008	case control	63,081	hip	1.09 (0.88–1.34)
			any	0.99 (0.90–1.11)
Kaye and Jick [7], 2008	case control	12,021	hip	0.90 (0.72–1.13)
Roux et al. [8], 2009	cohort	1,211	hip	0.55 (0.07–4.44)
			spine	3.10 (1.14–8.44)
			any	0.79 (0.41–1.52)
de Vries et al. [9], 2009	cohort	234,144	hip	1.22 (1.09–1.36)
			spine	1.40 (1.11–1.77)
			wrist	1.09 (0.99–1.21)
Gray et al. [10], 2010	cohort	161,806	hip	1.00 (0.71–1.40)
			spine	1.47 (1.18–1.82)
			any	1.25 (1.15–1.36)
Corley et al. [11], 2010	case control	164,223	hip	1.30 (1.21–1.39)
Pouwels et al. [12], 2011	case control	33,104	hip	1.20 (1.04–1.40)
Chiu et al. [13], 2010	case control	2,482	hip	1.09 (1.35–2.11)
Mello et al. [14], 2012	cohort	9,437	any	3.87 (2.46–6.08)
Khalili et al. [15], 2012	cohort	79,899	hip	1.35 (1.13–1.62)

founding factors and bias cannot be appropriately excluded. To exclude confounding factors, research that explores the possible mechanisms of PPI-induced bone fracture will be necessary. In addition, a prospective randomized controlled study will also be important to exclude possible bias and clarify the true risk of long-term PPI administration in regard to different types of bone fractures.

Impaired Iron Absorption and Iron Deficiency Anemia

For effective absorption of nonheme iron from duodenal mucosa, iron must be converted from a nonabsorbable to an absorbable ferrous form by gastric acid. A severe hypochlorhydric state has been reported to be associated with iron deficiency and microcytic anemia. When gastric acid secretion is inhibited by chronic administration of PPIs, effective reduction of iron from Fe^{3+} to Fe^{2+} may be inhibited and iron absorption might be reduced. Therefore, long-term administration of PPIs may cause iron-deficient microcytic and hypochromic anemia.

A retrospective cohort study found statistically significant decreases in hemoglobin, hematocrit, and mean corpuscular volume in patients treated with PPIs in comparison with nontreated cases [22]. However, differences for those parameters observed over a 2-year observational period were small and prone to confounding factors because of the retrospective nature of the study. An important issue to consider is the presence of a potent duodenal mucosal regulatory mechanism of iron absorption that is under the control of body iron storage. When this regulatory mechanism is damaged, as in cases with hereditary hemochromatosis, PPI administration is reported to inhibit absorption of nonheme iron from meals [23]. However, long-term PPI administration has not been proven to decrease iron storage or induce microcytic anemia in subjects with an intact regulatory mechanism, though conflicting reports have been presented [22, 24, 25].

Impaired Vitamin B_{12} Absorption

Vitamin B_{12} is an important coenzyme for methylmalonyl CoA isomerization to succinyl CoA and enzymatic conversion of homocysteine to methionine, as well as for normal production of DNA and normal hematopoiesis. Thus, vitamin B_{12} deficiency is accompanied by increased plasma methylmalonic acid and homocysteine, and by a type of macrocytic anemia called pernicious anemia. Dietary vitamin B_{12} is tightly bound to protein consumed with a meal and requires gastric acid to be released. The released vitamin B_{12} first binds to salivary R proteins in the stomach and then to intrinsic factors in the small intestine. Thus, gastric acid secretion is necessary for the release of free vitamin B_{12} and its absorption. Free vitamin B_{12} bound with intrinsic factors moves down to the terminal ileum and is absorbed there. Secretion of these intrinsic factors from parietal cells is reported to be not inhibited by PPI administration. Therefore, only PPI-induced inhibition of gastric acid secretion is related to decreased vitamin B_{12} absorption by inhibiting the release of free vitamin B_{12}.

Observational studies have shown decreased serum vitamin B_{12} levels in long-term users of PPIs [26–28]. In addition, measurements of serum methylmalonic acid and homocysteine have frequently revealed subclinical vitamin B_{12} deficiency in PPI users. On the other hand, these observational studies have several limitations, such as small size, open-label characteristics, and lack of a placebo control group, which are necessary to exclude confounding factors. In addition, some observational studies failed to detect vitamin B_{12} deficiency even in long-term PPI users [24, 29]. Since all reported studies were observational and nonrandomized, prospective randomized studies are necessary to determine whether PPI administration causes vitamin B_{12} deficiency, which is clinically important. Vitamin B_{12} deficiency leads to neurological disorders such as subacute degeneration of the spinal cord, gait disturbance, depression, and dementia, though they can be reversed if appropriately treated. Therefore, it is important to periodically check serum vitamin B_{12}, methylmalonic acid, and homocysteine levels to detect subclinical vitamin B_{12} deficiency in patients treated with long-term PPI administration, especially elderly who are vulnerable to vitamin B_{12} deficiency, even if the risk may not be great.

Decreased Bactericidal Effect of Gastric Acid

A majority of pathogenic bacteria cannot survive in highly acidic conditions. Therefore, enteropathogenic bacteria such as *Salmonella*, *Campylobacter*, and *Clostridium difficile* ingested with food can be killed in the stomach. Full-dose PPI administration raises intragastric pH above 4.0 for over 50% of the time period that day; thus, PPIs may be related to increased enteric bacterial infections. The relationship of PPI administration with enteric bacterial infection is described in the chapter by Nakamura et al. [this vol., pp. 92–101].

PPI administration may also increase the population of various types of bacteria in the stomach and distal esophagus, where bacterial colonization is prevented by physiological gastric acid reflux. Increased bacterial colonization in those locations may be a pathogenetic factor for bacterial pneumonia through aspiration of refluxed contaminated esophageal contents.

Pneumonia

In 2004, a case-control study from the Netherlands that investigated the relationship between pneumonia and acid-suppressive drugs was reported. Patients who had pneumonia were shown to have a higher chance of having been treated with PPIs [30] and the odds ratio (OR) for development of pneumonia following PPI administration was found to be approximately 2. Thereafter, two similar case control studies from Denmark and the UK were reported [31, 32]. The Denmark investigators found that administration of PPIs raised the risk of pneumonia, and the OR was 1.5. In addition, they reported that the OR was higher (5.0) during the first week after the start of PPI administration, but decreased to 1.3 thereafter. In 2008, the UK investigators reported an OR of 6.53 for pneumonia during the first 2 administration days, then 3.79 during the first and 3.21 during the first and second administration weeks, with no statistically significant risk noted after the third administration week.

These case control studies are prone to a number of confounding factors and biases, and their reliability is limited. The finding of decreasing risk during continuous long-term administration is difficult to explain since the acid-suppressing effect of PPIs gradually becomes stronger over several days before showing the strongest acid inhibition. If pneumonia is related to PPI-induced acid inhibition, it is reasonable to expect that the risk of pneumonia will increase during long-term administration. Such a discrepancy with the expected results raises concerns regarding the reliability of the reported risk. In 2009, a cohort study investigated hospital-acquired pneumonia in relationship to PPIs and reported an OR of 1.3 for hospital-acquired pneumonia [33]. That study was conducted in a hospital where the mean inpatient stay was as short as 5 days. It is difficult to accept that PPI administration caused clinically meaningful pneumonia within such a short period or that it could be effectively treated prior to safe discharge of the affected patients. On the other hand, PPIs were reported to not be a risk factor for pneumonia in a recent large-scale case control study that included 1,125 pneumonia cases and 2,235 controls [34].

Table 3. PPI administration and reported risk of pneumonia

Study	Study design	n	Drug	RR for pneumonia (95% CI)
Laheij et al. [38], 2003	cohort	405	PPIs/H_2-RAs	18.2 (2–158)
Laheij et al. [30], 2004	case control	5,165	PPIs	1.89 (1.36–2.62)
Gulmez et al. [31], 2007	case control	41,818	PPIs	1.50 (1.31–1.72)
Sarkar et al. [32], 2008	case control	879,947	PPIs	2.05 (1.96–2.15)
Myles et al. [39], 2009	case control	25,883	PPIs	1.55 (1.36–1.77)
Rodriguez et al. [40], 2009	case control	17,290	PPIs	1.16 (1.03–1.31)
Marciniak et al. [41], 2009	case control	72	PPIs	1.80 (0.50–6.80)
Herzig et al. [33], 2009	cohort	63,878	PPIs	1.30 (1.10–1.40)
Roughead et al. [42], 2009	cohort	185,533	PPIs	1.16 (1.11–1.22)
Dublin et al. [34], 2010	case control	3,360	PPIs/H_2-RA	1.13 (0.88–1.45)
Eurich et al. [43], 2010	case control	2,724	PPIs	1.45 (1.03–2.04)
Redelmeier et al. [44], 2010	cohort	593,265	PPIs/H_2-RA	1.02 (0.96–1.09)
Gau et al. [45], 2010	case control	1,146	PPIs	1.18 (0.80–1.74)
Meijvis et al. [46], 2011	case control	2,150	PPIs	3.10 (1.40–7.10)
Hermos et al. [47], 2012	case control	16,984	PPIs	1.29 (1.15–1.45)

H_2-RA = H_2-receptor antagonist.

Because of this inconsistency among research results, several meta-analyses have been performed. According to a recently published meta-analysis that included nine observational studies, the OR for current use of PPIs and pneumonia was 1.39 [35]. Shorter durations of PPI treatment of less than 30 days and higher doses have been reported to be associated with higher risks, with an OR of 1.65 and 1.50, respectively. Interestingly, when PPIs were given for longer than 180 days, the risk of pneumonia disappeared. This meta-analysis clearly summarized previous reports, and clarified the large heterogeneity of the data provided by the original studies as well as strong limitations of the observational protocols used.

Pooled and meta-analyses of results obtained in prospective randomized double-blind placebo-controlled studies have higher reliability and fewer confounding factors. These studies have consistently shown that administration of PPIs does not significantly increase the risk of respiratory infection [36, 37]. In a pooled analysis that included 31 randomized trials with a total of 3,358 patients given a placebo and 6,534 administered esomeprazole, the risk ratio (RR) for esomeprazole as compared to the placebo was 0.93 for all respiratory tract infection, 0.85 for symptoms suggesting respiratory tract infection, 0.92 for lower respiratory tract infection, and 0.94 for pneumonia [36]. These results suggest the possible inclusion of confounding factors and biases in other previously published observational studies (table 3). Therefore, the risk of PPI administration for pneumonia seems to be small. A practical approach is vigilance about the possible occurrence of pneumonia during the first 2 months after beginning PPI administration.

Decreased Acid Secretion and Altered Absorption of Coadministered Drugs

The presence of gastric acid is known to influence the degradation, solubility, and absorption of orally administered drugs. Acid-unstable drugs are degraded by gastric acid, while some drugs require acidification for effective solubilization and following absorption. Therefore, PPI-induced high-grade inhibition of gastric acid secretion can alter the bioavailability of coadministered drugs [48].

Increased Absorption of Digoxin, Methyldigoxin, Nifedipine, and Alendronate

Digoxin and methyldigoxin are partially degraded by gastric acid to the aglycone digoxigenin. Acid inhibition is considered to decrease this degradation and augment the bioavailability of digoxin and methyldigoxin. The stability of nifedipine is reported to be pH dependent and it is partially degraded in a strongly acidic condition such as gastric juice. Therefore, acid suppression is also expected to increase the stability of ingested nifedipine and increase its bioavailability. Alendronate is increasingly being used for treatment of osteoporosis. Its solubility in an aqueous solution changes with the pH of the solution, as its solubility is low in an acidic condition and increases with neutralization of the solution. Therefore, PPI administration increases the solubility of alendronate and increases its absorption.

Digoxin, methyldigoxin, and nifedipine are potentially toxic drugs, with their own appropriate therapeutic plasma concentrations. Therefore, care is needed to recheck the therapeutic effects of these drugs to detect possible overdose when PPIs are coadministered, though the clinical importance of such interactions has not been fully clarified.

Decreased Absorption of Atazanavir, Itraconazole, Ketoconazole, Cefpodoxime, Enoxacin, Gefitinib, Dipyridamole, and Aspirin

The HIV protease inhibitor atazanavir, antifungal agents itraconazole and ketoconazole, antibiotics cefpodoxime and enoxacin, tyrosine kinase inhibitor of epidermal growth factor receptors gefitinib, and antithrombotic agent dipyridamole are weak basic compounds, and their solubility in an acidic aqueous solution is higher. In neutralized gastric juice following PPI administration, their dissociation process in gastric juice is delayed and absorption from the intestine is expected to decrease. Such decreased absorption of medicine given for infections by pathogenic microorganisms can lead to critical results in certain conditions. Similarly, decreased absorption of an antithrombotic agent may cause increased risk of thromboembolic events. Recently, PPI administration was reported to decrease the antithrombotic effect of low-dose aspirin by decreasing its absorption. In a rat study, PPI delayed the gastrointestinal absorption of aspirin [49]. However, in human studies, the interactions of aspirin with PPIs were not consistent and the evidence provided in those reports is inadequate [50–53]. Thus, more reliable studies are anticipated.

Interaction of Proton Pump Inhibitors at CYP2C19 with Coadministered Drugs

Many kinds of drugs, such as clopidogrel, warfarin, tacrolimus, and diazepam, are activated or degraded by the hepatic cytochrome p450 enzyme CYP2C19. All PPIs are at least partially degraded by CYP2C19, though they differ. It has been shown that PPIs that require CYP2C19 for degradation are involved with competitive inhibition of the enzyme, which is necessary for the metabolism of coadministered drugs.

Clopidogrel

Clopidogrel is widely used as an antithrombotic drug for primary and secondary prevention of cerebrovascular and cardiovascular diseases. Following gastrointestinal absorption, the drug requires activation by hepatic cytochrome p450 enzymes including CYP2C19 for its pharmacological antithrombotic effect. Ex vivo studies using platelets obtained from patients taking clopidogrel alone and those taking clopidogrel with PPIs have shown that the latter coadministration decreased its antithrombotic effect [54, 55]. A number of clinical studies that followed those ex vivo studies found that PPI administration decreased the antithrombotic effect of clopidogrel and increased the risk of major cardiovascular events, including acute myocardial infraction, though some authors did not find a significantly increased risk associated with coadministration with PPIs [56, 57]. Nearly all of the reported studies are retrospective and observational in nature, and the findings are susceptible to influence by confounding factors. Indeed, a study that reanalyzed the results of previously published randomized controlled studies showed that PPI coadministration with clopidogrel did not elevate cardiovascular risk [58]. Data obtained from prospective randomized controlled studies are expected to have a lower level of influence from such biases.

As for the possible interaction of clopidogrel and PPIs, only two prospective randomized controlled studies have been presented [59, 60]. In one of those, 3,761 patients who underwent dual-antiplatelet therapy including aspirin and clopidogrel for acute coronary syndrome or received placement of a coronary stent were enrolled. They were randomly assigned to omeprazole and placebo groups, and gastrointestinal and cardiovascular events were compared. As expected, omeprazole administration significantly decreased upper gastrointestinal bleeding complications (hazard ratio: 0.13, range: 0.03–0.56). On the other hand, omeprazole administration did not augment cardiovascular events (hazard ratio: 0.99, 0.68–1.44). Therefore, even though the study was terminated prior to the planned endpoint, PPIs may be considered to have a scant inhibitory effect on the antithrombotic effect of clopidogrel in clinical situations, even if there is some ex vivo effect.

The other prospective randomized controlled study obtained nearly the same results, though the number of enrolled patients was smaller [60]. Furuta et al. [61] recently compared the different effects of PPIs and genotypes of CYP2C19 on clopidogrel-induced inhibition of platelet aggregation. They found that the effect of clopidogrel on platelet aggregation was weaker in decreased metabolizers of CYP2C19 as compared to rapid metabolizers. They also found that the effects of PPIs on clopidogrel-induced inhibition of platelet aggregation were much lower than those of genotypes of CYP2C19. Although these results do not completely eliminate the cardiovascular risk of PPIs in clopidogrel-treated patients, the risk seems to be smaller than previously proposed. The current practical approach is risk stratification of PPI administration and appropriate selection of patients who receive PPIs.

Warfarin

Warfarin is a racemic compound composed of S- and R-isomers. S-warfarin, reported to be five times stronger than R-isomers, is mainly degraded by the hepatic cytochrome p450 enzyme CYP2C9. On the other hand, R-isomers are degraded by cytochrome p450 enzymes including CYP2C19. PPIs may interfere with R-warfarin degradation and may raise its plasma concentration in treated patients, though the effects on the more potent S-isomer of warfarin are expected to be less significant.

The antithrombotic effect of warfarin is strongly influenced by various drug-related and unrelated factors [62]. Antibiotics, antifungal agents, and selective serotonin reuptake inhibitors are well-known drugs that augment the pharmacological effect of warfarin. As a member of this group of medications, PPIs must be considered, and patients treated by coadministration of warfarin and a PPI should be periodically monitored for antithrombotic effects.

Diazepam, Phenytoin, Cilostazol, and Tacrolimus

Diazepam, phenytoin, and cilostazol are mainly degraded by the hepatic enzyme CYP2C19. Therefore, when PPIs mainly degraded by CYP2C19 are coadministered with these drugs, their plasma concentrations are considered to rise because of the degradation enzyme competition. Indeed, PPIs are reported to elevate the plasma level of diazepam, phenytoin, and cilostazol via this mechanism. PPIs are also reported to elevate plasma tacrolimus concentration by interacting with tacrolimus metabolism at CYP3As, especially when these drugs are administered to CYP2C19 poor metabolizers [63, 64]. These drugs are potentially dangerous when the plasma concentration is elevated above the therapeutic level. Careless administration of PPIs with such medications may elevate their plasma concentrations and cause unintended toxicity.

Other Possible Important Systemic Side Effects

In addition to the side effects mentioned earlier, many kinds of rare but serious systemic side effects of PPIs have been reported, including anaphylaxis, pancytopenia, agranulocytosis, fulminant hepatitis, toxic epidermal necrolysis, Stevens-

Johnson syndrome, interstitial nephritis, interstitial pneumonitis, rhabdomyolysis, and hypomagnesemia [65–67]. These have also been reported in patients administered a variety of drugs, though the pathogeneses are not well understood. Since the side effects are possibly allergic reactions, the pathogenetic mechanisms, different from the other side effects mentioned above, may not be related to the pharmacological action of PPIs. Therefore, it is considered difficult to predict the occurrence of such adverse events. Careful follow-up of patients treated with PPIs is necessary as with those treated with other drugs, even though the side effects of PPI administration are not higher in frequency than those of other widely used medications.

Conclusion

PPIs can cause various systemic side effects, some of which are related to their pharmacological action and predictable, while those that are related to allergic reactions are difficult to predict. Since PPIs are widely used in long-term treatment with other drugs, side effects and possible interactions with coadministered drugs must be carefully watched, even if the incidence is low. Future studies will be necessary to solve confusion and clarify the clinical importance of reported side effects of PPIs, while the practical method at present is to carefully consider the indications for their administration.

References

1 O'Connell MB, Madden DM, Murray AM, Heaney RP, Kerzner LJ: Effects of proton pump inhibitors on calcium carbonate absorption in women: a randomized crossover trial. Am J Med 2005;118:778–781.

2 Eastell R, Vieira NE, Yergey AL, Wahner HW, Silverstein MN, Kumar R, Riggs BL: Pernicious anaemia as a risk factor for osteoporosis. Clin Sci 1992;82:681–685.

3 Yang YX, Lewis JD, Epstein S, Metz DC: Long-term proton pump inhibitor therapy and risk of hip fracture. JAMA 2006;296:2947–2953.

4 Vestergaard P, Rejnmark L, Mosekilde L: Proton pump inhibitors, histamine H_2 receptor antagonists, and other antacid medications and the risk of fracture. Calcif Tissue Int 2006;79:76–83.

5 Yu EW, Blackwell T, Ensrud KE, Hillier TA, Lane NE, Orwoll E, Bauer DC: Acid-suppressive medications and risk of bone loss and fracture in older adults. Calcif Tissue Int 2008;83:251–259.

6 Targownik LE, Lix LM, Metge CJ, Prior HJ, Leung S, Leslie WD: Use of proton pump inhibitors and risk of osteoporosis-related fractures. CMAJ 2008;179:319–326.

7 Kaye JA, Jick H: Proton pump inhibitor use and risk of hip fractures in patients without major risk factors. Pharmacotherapy 2008;28:951–959.

8 Roux C, Briot K, Gossec L, Kolta S, Blenk T, Felsenberg D, Reid DM, Eastell R, Glüer CC: Increase in vertebral fracture risk in postmenopausal women using omeprazole. Calcif Tissue Int 2009;84:13–19.

9 de Vries F, Cooper AL, Cockle SM, van Staa TP, Cooper C: Fracture risk in patients receiving acid-suppressant medication alone and in combination with bisphosphonates. Osteoporos Int 2009;20:1989–1998.

10 Gray SL, LaCroix AZ, Larson J, Robbins J, Cauley JA, Manson JE, Chen Z: Proton pump inhibitor use, hip fracture, and change in bone mineral density in postmenopausal women: results from the Women's Health Initiative. Arch Intern Med 2010;170:765–771.

11 Corley DA, Kubo A, Zhao W, Quesenberry C: Proton pump inhibitors and histamine-2 receptor antagonists are associated with hip fractures among at-risk patients. Gastroenterology 2010;139:93–101.
12 Pouwels S, Lalmohamed A, Souverein P, Cooper C, Veldt BJ, Leufkens HG, de Boer A, van Staa T, de Vries F: Use of proton pump inhibitors and risk of hip/femur fracture: a population-based case-control study. Osteoporos Int 2011;22:903–910.
13 Chiu HF, Huang YW, Chang CC, Yang CY: Use of proton pump inhibitors increased the risk of hip fracture: a population-based case-control study. Pharmacoepidemiol Drug Saf 2010;19:1131–1136.
14 Mello M, Weideman RA, Little BB, Weideman MW, Cryer B, Brown GR: Proton pump inhibitors increase the incidence of bone fractures in hepatitis C patients. Dig Dis Sci 2012;57:2416–2422.
15 Khalili H, Huang ES, Jacobson BC, Camargo CA Jr, Feskanich D, Chan AT: Use of proton pump inhibitors and risk of hip fracture in relation to dietary and lifestyle factors: a prospective cohort study. BMJ 2012;344:e372.
16 Yu EW, Bauer SR, Bain PA, Bauer DC: Proton pump inhibitors and risk of fractures: a meta-analysis of 11 international studies. Am J Med 2011;124:519–526.
17 Ngamruengphong S, Leontiadis GI, Radhi S, Dentino A, Nugent K: Proton pump inhibitors and risk of fracture: a systematic review and meta-analysis of observational studies. Am J Gastroenterol 2011;106: 1209–1218.
18 Eom CS, Park SM, Myung SK, Yun JM, Ahn JS: Use of acid-suppressive drugs and risk of fracture: a meta-analysis of observational studies. Ann Fam Med 2011;9:257–267.
19 Kwok CS, Yeong JK, Loke YK: Meta-analysis: risk of fractures with acid-suppressing medication. Bone 2011;48:768–776.
20 Targownik LE, Lix LM, Leung S, Leslie WD: Proton-pump inhibitor use is not associated with osteoporosis or accelerated bone mineral density loss. Gastroenterology 2010;138:896–904.
21 Targownik LE, Leslie WD, Davison KS, Goltzman D, Jamal SA, Kreiger N, Josse RG, Kaiser SM, Kovacs CS, Prior JC, Zhou W, the CaMos Research Group: The relationship between proton pump inhibitor use and longitudinal change in bone mineral density: a population-based study from the Canadian Multicentre Osteoporosis Study (CaMos). Am J Gastroenterol 2012;107:1361–1369.
22 Sarzynski E, Puttarajappa C, Xie Y, Grover M, Laird-Fick H: Association between proton pump inhibitor use and anemia: a retrospective cohort study. Dig Dis Sci 2011;56:2349–2353.
23 Hutchinson C, Geissler CA, Powell JJ, Bomford A: Proton pump inhibitors suppress absorption of dietary non-haem iron in hereditary haemochromatosis. Gut 2007;56:1291–1295.
24 Koop H, Bachem MG: Serum iron, ferritin, and vitamin B_{12} during prolonged omeprazole therapy. J Clin Gastroenterol 1992;14:288–292.
25 Stewart CA, Termanini B, Sutliff VE, Serrano J, Yu F, Gibril F, Jensen RT: Iron absorption in patients with Zollinger-Ellison syndrome treated with long-term gastric acid antisecretory therapy. Aliment Pharmacol Ther 1998;12:83–98.
26 Hirschowitz BI, Worthington J, Mohnen J: Vitamin B_{12} deficiency in hypersecretors during long-term acid suppression with proton pump inhibitors. Aliment Pharmacol Ther 2008;27:1110–1121.
27 Valuck RJ, Ruscin JM: A case-control study on adverse effects: H_2 blocker or proton pump inhibitor use and risk of vitamin B_{12} deficiency in older adults. J Clin Epidemiol 2004;57:422–428.
28 Rozgony NR, Fang C, Kuczmarski MF, Bob H: Vitamin B(12) deficiency is linked with long-term use of proton pump inhibitors in institutionalized older adults: could a cyanocobalamin nasal spray be beneficial? J Nutr Elder 2010;29:87–99.
29 den Elzen WP, Groeneveld Y, de Ruijter W, Souverijn JH, le Cessie S, Assendelft WJ, Gussekloo J: Long-term use of proton pump inhibitors and vitamin B_{12} status in elderly individuals. Aliment Pharmacol Ther 2008;27:491–497.
30 Laheij RJ, Sturkenboom MC, Hassing RJ, Dieleman J, Stricker BH, Jansen JB: Risk of community-acquired pneumonia and use of gastric acid-suppressive drugs. JAMA 2004;292:1955–1960.
31 Gulmez SE, Holm A, Frederiksen H, Jensen TG, Pedersen C, Hallas J: Use of proton pump inhibitors and the risk of community-acquired pneumonia: a population-based case-control study. Arch Intern Med 2007;167:950–955.
32 Sarkar M, Hennessy S, Yang YX: Proton-pump inhibitor use and the risk for community-acquired pneumonia. Ann Intern Med 2008;149:391–398.
33 Herzig SJ, Howell MD, Ngo LH, Marcantonio ER: Acid-suppressive medication use and the risk for hospital-acquired pneumonia. JAMA 2009;301:2120–2128.
34 Dublin S, Walker RL, Jackson ML, Nelson JC, Weiss NS, Jackson LA: Use of proton pump inhibitors and H_2 blockers and risk of pneumonia in older adults: a population-based case-control study. Pharmacoepidemiol Drug Saf 2010;19:792–802.
35 Giuliano C, Wilhelm SM, Kale-Pradhan PB: Are proton pump inhibitors associated with the development of community-acquired pneumonia? A meta-analysis. Expert Rev Clin Pharmacol 2012;5:337–344.
36 Estborn L, Joelson S: Occurrence of community-acquired respiratory tract infection in patients receiving esomeprazole: retrospective analysis of adverse events in 31 clinical trials. Drug Saf 2008;31:627–636.

37 Sultan N, Nazareno J, Gregor J: Association between proton pump inhibitors and respiratory infections: a systematic review and meta-analysis of clinical trials. Can J Gastroenterol 2008;22:761–766.
38 Laheij RJ, Van Ijzendoorn MC, Janssen MJ, Jansen JB: Gastric acid-suppressive therapy and community-acquired respiratory infections. Aliment Pharmacol Ther 2003;18:847–851.
39 Myles PR, Hubbard RB, McKeever TM, Pogson Z, Smith CJ, Gibson JE: Risk of community-acquired pneumonia and the use of statins, ace inhibitors and gastric acid suppressants: a population-based case-control study. Pharmacoepidemiol Drug Saf 2009; 18:269–275.
40 Rodríguez LA, Ruigómez A, Wallander MA, Johansson S: Acid-suppressive drugs and community-acquired pneumonia. Epidemiology 2009;20:800–806.
41 Marciniak C, Korutz AW, Lin E, Roth E, Welty L, Lovell L: Examination of selected clinical factors and medication use as risk factors for pneumonia during stroke rehabilitation: a case-control study. Am J Phys Med Rehabil 2009;88:30–38.
42 Roughead EE, Ramsay EN, Pratt NL, Ryan P, Gilbert AL: Proton-pump inhibitors and the risk of antibiotic use and hospitalisation for pneumonia. Med J Aust 2009;190:114–116.
43 Eurich DT, Sadowski CA, Simpson SH, Marrie TJ, Majumdar SR: Recurrent community-acquired pneumonia in patients starting acid-suppressing drugs. Am J Med 2010;123:47–53.
44 Redelmeier DA, McAlister FA, Kandel CE, Lu H, Daneman N: Postoperative pneumonia in elderly patients receiving acid suppressants: a retrospective cohort analysis. BMJ 2010;340:c2608.
45 Gau JT, Acharya U, Khan S, Heh V, Mody L, Kao TC: Pharmacotherapy and the risk for community-acquired pneumonia. BMC Geriatr 2010;10:45.
46 Meijvis SC, Cornips MC, Voorn GP, Souverein PC, Endeman H, Biesma DH, Leufkens HG, van de Garde EM: Microbial evaluation of proton-pump inhibitors and the risk of pneumonia. Eur Respir J 2011;38:1165–1172.
47 Hermos JA, Young MM, Fonda JR, Gagnon DR, Fiore LD, Lawler EV: Risk of community-acquired pneumonia in veteran patients to whom proton pump inhibitors were dispensed. Clin Infect Dis 2012;54:33–42.
48 Lahner E, Annibale B, Delle Fave G: Systematic review: impaired drug absorption related to the co-administration of antisecretory therapy. Aliment Pharmacol Ther 2009;29:1219–1229.
49 Giraud MN, Sanduja SK, Felder TB, Illich PA, Dial EJ, Lichtenberger LM: Effect of omeprazole on the bioavailability of unmodified and phospholipid-complexed aspirin in rats. Aliment Pharmacol Ther 1997;11:899–906.
50 Würtz M, Grove EL, Kristensen SD, Hvas AM: The antiplatelet effect of aspirin is reduced by proton pump inhibitors in patients with coronary artery disease. Heart 2010;96:368–371.
51 Kasprzak M, Koziński M, Bielis L, Boinska J, Plazuk W, Marciniak A, Budzyński J, Siller-Matula J, Rość D, Kubica J: Pantoprazole may enhance antiplatelet effect of enteric-coated aspirin in patients with acute coronary syndrome. Cardiol J 2009;16:535–544.
52 Iñarrea P, Esteva F, Cornudella R, Lanas A: Omeprazole does not interfere with the antiplatelet effect of low-dose aspirin in man. Scand J Gastroenterol 2000; 35:242–246.
53 Charlot M, Grove EL, Hansen PR, Olesen JB, Ahlehoff O, Selmer C, Lindhardsen J, Madsen JK, Køber L, Torp-Pedersen C, Gislason GH: Proton pump inhibitor use and risk of adverse cardiovascular events in aspirin treated patients with first time myocardial infarction: nationwide propensity score matched study. BMJ 2011;342:d2690.
54 Gilard M, Arnaud B, Le Gal G, Abgrall JF, Boschat J: Influence of omeprazol on the antiplatelet action of clopidogrel associated to aspirin. J Thromb Haemost 2006;4:2508–2509.
55 Sibbing D, Morath T, Stegherr J, Braun S, Vogt W, Hadamitzky M, Schömig A, Kastrati A, von Beckerath N: Impact of proton pump inhibitors on the antiplatelet effects of clopidogrel. Thromb Haemost 2009;101:714–719.
56 Kwok CS, Loke YK: Meta-analysis: the effects of proton pump inhibitors on cardiovascular events and mortality in patients receiving clopidogrel. Aliment Pharmacol Ther 2010;31:810–823.
57 Disney BR, Watson RD, Blann AD, Lip GY, Anderson MR: Review article: proton pump inhibitors with clopidogrel – evidence for and against a clinically-important interaction. Aliment Pharmacol Ther 2011;33: 758–767.
58 O'Donoghue ML, Braunwald E, Antman EM, Murphy SA, Bates ER, Rozenman Y, Michelson AD, Hautvast RW, Ver Lee PN, Close SL, Shen L, Mega JL, Sabatine MS, Wiviott SD: Pharmacodynamic effect and clinical efficacy of clopidogrel and prasugrel with or without a proton-pump inhibitor: an analysis of two randomised trials. Lancet 2009;374:989–997.
59 Bhatt DL, Cryer BL, Contant CF, Cohen M, Lanas A, Schnitzer TJ, Shook TL, Lapuerta P, Goldsmith MA, Laine L, Scirica BM, Murphy SA, Cannon CP, COGENT Investigators: Clopidogrel with or without omeprazole in coronary artery disease. N Engl J Med 2010;363:1909–1917.
60 Hsu PI, Lai KH, Liu CP: Esomeprazole with clopidogrel reduces peptic ulcer recurrence, compared with clopidogrel alone, in patients with atherosclerosis. Gastroenterology 2011;140:791–798.

61 Furuta T, Iwaki T, Umemura K: Influences of different proton pump inhibitors on the anti-platelet function of clopidogrel in relation to CYP2C19 genotypes. Br J Clin Pharmacol 2010;70:383–392.
62 Baillargeon J, Holmes HM, Lin YL, Raji MA, Sharma G, Kuo YF: Concurrent use of warfarin and antibiotics and the risk of bleeding in older adults. Am J Med 2012;125:183–189.
63 Itagaki F, Homma M, Yuzawa K, Nishimura M, Naito S, Ueda N, Ohkohchi N, Kohda Y: Effect of lansoprazole and rabeprazole on tacrolimus pharmacokinetics in healthy volunteers with CYP2C19 mutations. J Pharm Pharmacol 2004;56:1055–1059.
64 Hosohata K, Masuda S, Ogura Y, Oike F, Takada Y, Katsura T, Uemoto S, Inui K: Interaction between tacrolimus and lansoprazole, but not rabeprazole in living-donor liver transplant patients with defects of CYP2C19 and CYP3A5. Drug Metab Pharmacokinet 2008;23:134–138.
65 Sierra F, Suarez M, Rey M, Vela MF: Systematic review: Proton pump inhibitor-associated acute interstitial nephritis. Aliment Pharmacol Ther 2007;26:545–553.
66 Hoorn EJ, van der Hoek J, de Man RA, Kuipers EJ, Bolwerk C, Zietse R: A case series of proton pump inhibitor-induced hypomagnesemia. Am J Kidney Dis 2010;56:112–116.
67 Gau JT, Yang YX, Chen R, Kao TC: Uses of proton pump inhibitors and hypomagnesemia. Pharmacoepidemiol Drug Saf 2012;21:553–559.

Yoshikazu Kinoshita, MD, PhD
Department of Gastroenterology and Hepatology
Shimane University School of Medicine
89-1, Enya, Izumo, Shimane 693-8501 (Japan)
E-Mail kinoshita@med.shimane-u.ac.jp

Author Index

Subject Index